FM 3-76 (FM 3-05.60)

Special Operations Aviation

October 2011

DISTRIBUTION RESTRICTION: Distribution authorized to U.S. Government agencies and their contractors only to protect technical or operational information from automatic dissemination under the International Exchange Program or by other means. This determination was made on 1 September 2011. Other requests for this document must be referred to Commander, United States Army John F. Kennedy Special Warfare Center and School, ATTN: AOJK-CDI-CIC-JA, 3004 Ardennes Street, Stop A, Fort Bragg, NC 28310-9610.

DESTRUCTION NOTICE: Destroy by any method that will prevent disclosure of contents or reconstruction of the document.

FOREIGN DISCLOSURE RESTRICTION (FD 6): This publication has been reviewed by the product developers in coordination with the United States Army John F. Kennedy Special Warfare Center and School foreign disclosure authority. This product is releasable to students from foreign countries on a case-by-case basis only.

Headquarters, Department of the Army

This publication is available at
Army Knowledge Online (www.us.army.mil) and
General Dennis J. Reimer Training and Doctrine
Digital Library at (www.train.army.mil).

*FM 3-76 (FM 3-05.60)

Field Manual
No. 3-76 (FM 3-05.60)

Headquarters
Department of the Army
Washington, DC, 28 October 2011

Special Operations Aviation

Contents

		Page
	PREFACE	v
Chapter 1	ORGANIZATION AND FUNCTIONS	1-1
	Organization	1-1
	Task	1-3
	ARSOF Core Activities	1-3
	Functions of the SOAR	1-4
	SOAR Responsibilities in Support of ARSOF Core Activities	1-6
	Principles of SOAR Employment	1-7
	Employment Considerations	1-7
Chapter 2	MISSION COMMAND	2-1
	The Joint Operation Planning and Military Decisionmaking Processes	2-1
	Participation During the Joint Operation Planning and Military Decisionmaking Processes	2-1
	Joint Planning Considerations	2-1
	Joint Special Operations Air Component Commander	2-3
	Special Operations Liaison Element	2-4
	Aircraft Considerations	2-4
	Airspace Command and Control	2-4
	Airspace Deconfliction	2-5
	Airspace Control Measures	2-5
	Planning Processes	2-6
	Prevention of Fratricide	2-8

Distribution Restriction: Distribution authorized to U.S. Government agencies and their contractors only to protect technical or operational information from automatic dissemination under the International Exchange Program or by other means. This determination was made on 1 September 2011. Other requests for this document must be referred to Commander, United States Army John F. Kennedy Special Warfare Center and School, ATTN: AOJK-CDI-CIC-JA, 3004 Ardennes Street, Stop A, Fort Bragg, NC 28310-9610.

Destruction Notice: Destroy by any method that will prevent disclosure of contents or reconstruction of the document.

Foreign Disclosure Restriction (FD 6): This publication has been reviewed by the product developers in coordination with the United States Army John F. Kennedy Special Warfare Center and School foreign disclosure authority. This product is releasable to students from foreign countries on a case-by-case basis only.

*This publication supersedes FM 3-05.60, 30 October 2007.

Contents

Chapter 3	**OPERATIONS**	**3-1**
	Force Deployment, Employment, and Redeployment	3-1
	Environmental Considerations	3-3
	Key Operational Considerations	3-4
Chapter 4	**INTELLIGENCE**	**4-1**
	Intelligence Requirements	4-1
	Intelligence Organization	4-1
	Nonorganic Intelligence Support	4-5
	Intelligence Preparation of the Operational Environment	4-6
	Unmanned Aircraft Systems	4-9
Chapter 5	**COMMUNICATIONS**	**5-1**
	Communications Elements	5-1
	Communications Capabilities and Equipment	5-1
	Concept of Employment	5-2
	Unmanned Aircraft System Connectivity	5-5
Chapter 6	**FIRES**	**6-1**
	Terminology	6-1
	Coordination and Planning	6-1
	Assets and Techniques	6-2
Chapter 7	**SUSTAINMENT**	**7-1**
	Introduction	7-1
	Contingency Planning	7-1
	Crisis Action Planning	7-1
	Logistics Support	7-2
	Force Health Protection Support	7-5
	Funding and Finance Support	7-6
	Engineer Support	7-6
	Forward Arming and Refueling Point Operations	7-6
	Logistics in Developed and Undeveloped Theaters	7-6
	Contracting	7-7
Appendix A	**SOAR FORMATS**	**A-1**
Appendix B	**AIRCRAFT CAPABILITIES**	**B-1**
	GLOSSARY	**Glossary-1**
	REFERENCES	**References-1**
	INDEX	**Index-1**

Figures

Figure 1-1. Army Special Operations Aviation Command organization 1-1
Figure 1-2. 160th Special Operations Aviation Regiment (Airborne) organization 1-2
Figure 2-1. The military decisionmaking process 2-2
Figure 2-2. Principles of war and joint operations 2-3
Figure 2-3. Typical JSOACC command relationships 2-3

Contents

Figure 2-4. Direct support relationship 2-4
Figure 2-5. Typical special operations liaison element organization 2-4
Figure 2-6. The special operations mission-planning process 2-7
Figure 2-7. Timeline phase, 106 hours to 72 hours 2-9
Figure 2-8. Timeline phase, 72 hours to 48 hours 2-9
Figure 2-9. Timeline phase, 48 hours to takeoff (H-hour) 2-10
Figure 4-1. Intelligence organization of the SOAR 4-2
Figure 4-2. Intelligence organization of the SOAR battalion 4-2
Figure 4-3. Shadow platoon organization 4-10
Figure 4-4. MQ-1C (Grey Eagle) 4-11
Figure 4-5. RQ-7B (Shadow) 4-11
Figure 4-6. RQ-11B (Raven) 4-11
Figure 4-7. Wasp Block III (overhead and side views) 4-12
Figure 5-1. SOAR communications 5-2
Figure 5-2. SOAR communications network architecture 5-4
Figure 5-3. SOAR communications systems connectivity 5-5
Figure 5-4. MQ-1C (Grey Eagle) connectivity 5-6
Figure 5-5. RQ-7B (Shadow) connectivity 5-7
Figure 5-6. RQ-11B (Raven) connectivity 5-8
Figure A-1. SOAR operation order format A-1
Figure A-2. Mission tasking letter format A-15
Figure A-3. Feasibility assessment format A-16
Figure A-4. Initial assessment format A-17
Figure A-5. TIP format for direct action and special reconnaissance missions A-19
Figure A-6. TIP format for foreign internal defense and unconventional warfare missions A-21
Figure A-7. Mission tasking package format A-22
Figure A-8. SOF plan of execution format A-23
Figure A-9. Infiltration and exfiltration plan of execution format A-26
Figure A-10. Special operations aviation call-for-fire format A-28
Figure B-1. MH-6M helicopter B-1
Figure B-2. MH-6M and AH-6M aircraft dimensions B-4
Figure B-3. MH-6M and AH-6M aircraft dimensions and turning radius B-4
Figure B-4. AH-6M helicopter B-5
Figure B-5. AH-6M plank system for aircraft weapons configurations B-7
Figure B-6. AH-6M weapons variations B-7
Figure B-7. AH-6M safety approach areas B-10
Figure B-8. MH-60L helicopter B-11
Figure B-9. MH-60L fast-rope insertion and extraction system bar B-12
Figure B-10. MH-60L defensive armed penetrator B-13
Figure B-11. Armament options for the MH-60L defensive armed penetrator ... B-14

Figure B-12. MH-60K helicopter B-18
Figure B-13. MH-60K M134 minigun window-mounted field of fire B-19
Figure B-14. MH-60K dimensions and turning radius B-22
Figure B-15. MH-60K dimensions for intertheater airlift preparation B-23
Figure B-16. MH-60K aircraft capabilities B-24
Figure B-17. MH-47G helicopter B-25

Tables

Table 4-1. Army special operations unmanned aircraft system platforms
    and payloads 4-9
Table 7-1. Water requirements for aircraft washing and engine flushing (gallons) 7-3
Table 7-2. Force health protection personnel authorizations for the SOAR 7-5
Table B-1. MH-6M aircraft capabilities B-3
Table B-2. AH-6M aircraft capabilities B-9
Table B-3. MH-60L aircraft capabilities B-17
Table B-4. MH-60K aircraft capabilities B-21
Table B-5. MH-47G external cargo hooks B-26
Table B-6. MH-47G aircraft capabilities B-28
Table B-7. SOAR aircraft capabilities matrix B-29

Preface

Field Manual (FM) 3-76, *Special Operations Aviation*, describes the core activities, capabilities, limitations, command and control (C2) relationships, employment principles, and operational considerations of Army special operations aviation (SOA). It delineates unique capabilities, limitations, and requirements when supporting a standing joint special operations task force (JSOTF) or a geographic combatant commander during a regional operation.

PURPOSE

FM 3-76 establishes doctrine for operational employment of aviation logistics and sustainment, and command and support relationships. The manual identifies requirements of the Special Operations Aviation Regiment (SOAR) in different levels of conflict and throughout the duration of operations.

SCOPE

This manual explains short- and long-term employment and execution concepts for the SOAR, including support, sustainment requirements, and relationships. The manual provides doctrinal guidance to the SOAR commander for employment of the SOAR in support of special operations (SO). The employment guidance considerations and the command and support responsibilities apply to the special operations forces (SOF) and conventional commanders and their staffs.

APPLICABILITY

This publication applies to the Active Army, the Army National Guard/Army National Guard of the United States, and the United States Army Reserve unless otherwise stated.

ADMINISTRATIVE INFORMATION

This manual is unclassified to ensure Armywide dissemination and to facilitate the integration of Army special operations forces (ARSOF) in the preparation and execution of campaigns and major operations. Unless this publication states otherwise, masculine nouns and pronouns do not refer exclusively to men. The proponent of this manual is the United States Army John F. Kennedy Special Warfare Center and School (USAJFKSWCS). Submit comments and recommended changes on DA Form 2028 (Recommended Changes to Publications and Blank Forms) directly to Commander, USAJFKSWCS, ATTN: AOJK-CDI-CIC-JA, 3004 Ardennes Street, Stop A, Fort Bragg, NC 28310-9610, or by electronic DA Form 2028.

This page intentionally left blank.

Chapter 1

Organization and Functions

ORGANIZATION

1-1. The United States (U.S.) Army's SOA is orchestrated from the Army Special Operations Aviation Command (ARSOAC) headquartered at Fort Bragg, North Carolina. The ARSOAC organizes, mans, trains, resources, and equips Army SOA units to provide responsive SOA support to SO. Additionally, the command serves as the United States Army Special Operations Command (USASOC) aviation staff proponent, and includes a technology applications program office, a flight detachment, a systems integration management office, a regimental organizational applications element, a special operations aviation training battalion, and the 160th Special Operations Aviation Regiment (Airborne) (SOAR[A]). (Figure 1-1).

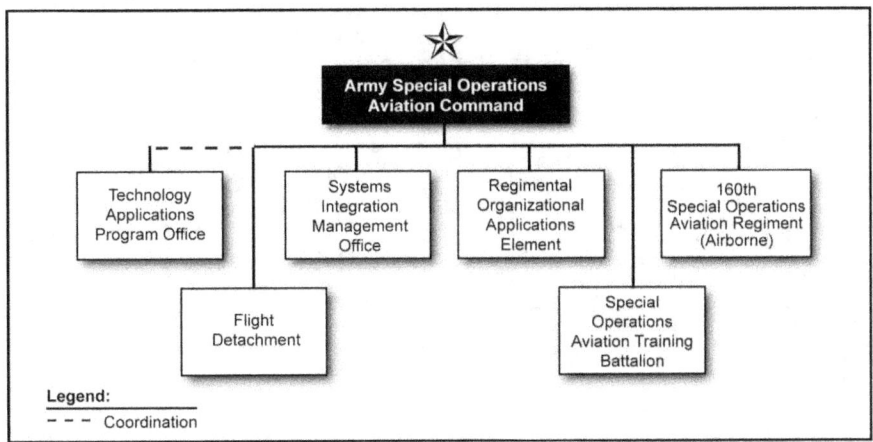

Figure 1-1. Army Special Operations Aviation Command organization

1-2. The technology applications program office uses streamlined acquisition procedures to rapidly procure and integrate nondevelopmental item equipment and systems for Army SOA, manages SOA's modifications and configuration control, provides logistics sustainment of SOA-peculiar equipment and systems, and, where applicable, transitions SOA-peculiar equipment and systems to conventional Army aircraft. The flight detachment provides responsive fixed- and rotary-wing training support to USASOC and provides key planner transport to the USASOC Commanding General and staff in support of contingency plans. The systems integration management office equips the Soldiers of the 160th SOAR(A) with the most capable rotary-wing aircraft in the world and facilitates the sustainment of highly modified and/or unique aircraft. The regimental organizational applications element develops and validates equipment; concepts; and tactics, techniques, and procedures to counter asymmetric threats in support of the United States Special Operations Command (USSOCOM), Department of Energy, and other government agencies. The training battalion conducts basic and advanced SOA air, ground, and aquatics training within the continental United States in order to produce basic mission-qualified crewmembers and

Chapter 1

support personnel for the 160th SOAR(A). The 160th SOAR(A) organizes, equips, trains, resources, and employs Army SOA forces worldwide in support of contingency missions and the warfighting commanders.

1-3. The Army's tactically operational SOA unit is the 160th SOAR[A] (Figure 1-2). It consists of a headquarters and headquarters company and four SOA battalions. Additionally, Table of Distribution and Allowance documents authorize a special operations aviation training company and a systems integration management office.

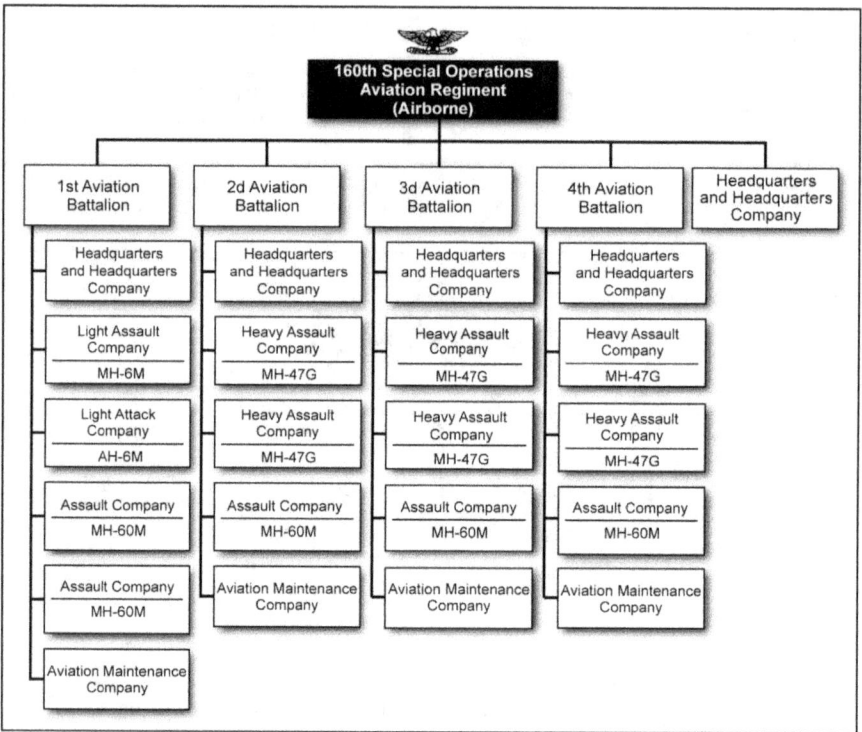

Figure 1-2. 160th Special Operations Aviation Regiment (Airborne) organization

1-4. The SOAR has SO rotary-wing aircraft, including the AH/MH-6M Little Bird, the MH-60L/K Blackhawk, and the MH-47G Chinook. SOAR units can conduct and support SO missions for the ARSOF commander or for the theater special operations command (TSOC). The SOAR can be task-organized based on expected missions, the requirements of the units being supported, the environmental conditions in the theater of operations, and sustainment requirements. The SOAR task-organizes around one of the SOA battalions. With proper personnel and equipment augmentation, the SOAR battalion commander and his staff could also serve as a joint special operations air component commander (JSOACC). When two or more battalions are required in the theater, the regimental commander could serve as the JSOACC.

1-5. The SOAR supports other SOF units by conducting special air operations in all operational environments. The specially organized, trained, and equipped aviation units give the joint force special operations component commander (JFSOCC) the capability to infiltrate, resupply, and exfiltrate SOF elements engaged in all core activities, missions, and environments.

TASK

1-6. The SOAR is tasked to conduct and support special air operations by clandestinely penetrating hostile and denied airspace. SOAR units can operate in harsh environments and across the full spectrum of operations. They also support SOF in conducting joint, combined, interagency, liaison, and coordination activities in support of the USSOCOM commander and the geographic combatant commander's concept of operations. The participation of the SOAR in the ARSOF core activities varies based upon the type of conflict, the environment, and the scope of the operation. As a component subordinate unit of the USASOC, the SOAR organizes, equips, trains, validates, sustains, and employs assigned aviation units for USASOC missions.

ARSOF CORE ACTIVITIES

1-7. The SOAR can support SOF in all of the core activities. The SOAR conducts these core activities predominantly in a joint environment and may support the U.S. military conventional forces, multinational forces, or other agencies. The following paragraphs discuss each of these core activities.

UNCONVENTIONAL WARFARE

1-8. In unconventional warfare, the SOAR's primary contributions are to infiltrate SOF into the joint special operations area (JSOA) or into a landing zone nearby where SOF can move into their JSOA. SOAR may also be tasked for resupply or exfiltration of SOF teams performing these missions.

FOREIGN INTERNAL DEFENSE

1-9. The SOAR assists foreign internal defense (FID) operations by providing aviation assets to supported multinational SOF requiring the SOAR's skills and equipment. It normally assists only in the movement of host-nation SOF to conduct surgical operations, such as precision application of fire with minimal or no collateral damage. In a FID environment, general aviation operations are normally inappropriate for SOAR missions. However, the SOAR may conduct or support tactical operations in support of FID operations. The objective of tactical operations in FID is to provide a secure environment in which balanced development can occur. Tactical operations should not be independent military operations aimed solely at destroying insurgent combat forces and their base areas. Tactical operations should be part of a synchronized effort to achieve the national strategic objectives of the host nation and the United States.

COUNTERINSURGENCY

1-10. As in FID, the SOAR assists counterinsurgency operations by providing aviation assets to supported multinational SOF requiring the SOAR's skills and equipment.

SECURITY FORCE ASSISTANCE

1-11. As in FID and counterinsurgency, the SOAR assists security force assistance operations by providing aviation assets to supported multinational SOF requiring the SOAR's skills and equipment.

DIRECT ACTION

1-12. In a direct action role, the SOAR's primary contribution is to infiltrate SOF directly onto the objective or into a landing zone where SOF can move to their target. With armed helicopters, SOAR units provide close air support and close combat attack, and conduct deep-penetration, unilateral direct action SO missions. The SOAR can support direct action SO units as small as split Special Forces (SF) teams and as large as a Ranger battalion. It can also conduct complex battalion-level air assault raids and support C2, communications, intelligence, deception, and show-of-force operations.

Chapter 1

SPECIAL RECONNAISSANCE

1-13. Normal SOAR support for special reconnaissance operations is through infiltration, resupply, or exfiltration of SOF teams performing these missions.

COUNTERTERRORISM

1-14. When conducting counterterrorism operations, the SOAR's responsibilities are the same as during a direct action mission. In addition, the SOAR may be tasked to transport high-value targets from the target area to a predetermined site for transfer to other forces or other government agencies.

MILITARY INFORMATION SUPPORT OPERATIONS

1-15. Conventional assets or Air Force SOF normally support overt Military Information Support operations (MISO), to include aerial broadcast and leaflet delivery. However, the SOAR normally supports clandestine MISO activities. The use of the SOAR is advisable when broadcast and leaflet delivery requires penetration of nonpermissive airspace to reach the MISO target audiences. The SOAR may be able to perform overt MISO activities if it is the only aviation asset available. FM 3-05.30, *Psychological Operations*, includes more information on MISO.

CIVIL AFFAIRS OPERATIONS

1-16. The SOAR can provide freedom of maneuver for Civil Affairs operations and coordination within the host country. Generally, the SOAR is of limited use in support of Civil Affairs. Normally, the SOAR only supports Civil Affairs in support of an SF unit in a FID or unconventional warfare environment. As in MISO, the best use of the SOAR is in covert or clandestine missions when hostile nations or the target audience requires penetration of nonpermissive airspace. FM 3-05.40, *Civil Affairs Operations*, includes more details on Civil Affairs operations and civil-military operations.

SUPPORT TO COUNTERPROLIFERATION OF WEAPONS OF MASS DESTRUCTION

1-17. Support for counterproliferation of weapons of mass destruction encompasses the same operational techniques used for direct action, counterterrorism, and special reconnaissance missions. Additionally, the SOAR may be tasked to transport seized equipment or materiel to selected sites.

SUPPORT TO INFORMATION OPERATIONS

1-18. SOAR support of information operations cuts across the entire range of military operations, from passive defense to precision strike operations against key information nodes. The goal of the SOAR within information operations is to achieve information dominance at the right time, at the right place, and with the right weapons or resources to gain and maintain information dominance.

FUNCTIONS OF THE SOAR

1-19. The unique capabilities of the SOAR are a function of the quality, training, and education of its Soldiers, as well as the mission profiles those Soldiers must execute. The competitive selection process, coupled with technological training and education, produces a Soldier who is adaptable, mature, innovative, culturally aware, self-assured, and self-reliant. Policy decisionmakers, therefore, use the SOAR as a strategic and operational economy of force to expand the range of available military options.

1-20. SOAR units can operate as part of a special operations task force (SOTF) or a JSOTF. They give the ground commander a means to infiltrate, sustain, and extract SOF. To employ the SOAR properly, commanders must understand the characteristics of SO in general and the SOAR in particular.

1-21. The SOAR conducts SO to achieve military, political, economic, or informational objectives by generally unconventional means in hostile, denied, or politically sensitive areas. Decisionmakers may choose the SOAR option because it provides the broadest range of capabilities that have direct applicability in an increasing number of environments.

1-22. Political and military requirements frequently cast the SOAR into clandestine or low-visibility environments that require oversight at the national level. SOAR operations differ from conventional force operations by their degree of acceptable physical and political risk, their modes of employment, and their operational techniques. The SOAR allows the unified commander or the joint force commander to perform critical and sensitive small-unit missions.

1-23. Early use of ARSOF in an operation may prevent or contain a conflict that may conserve national resources. When conflict is imminent, the SOAR may support a variety of prehostility missions to signal U.S. determination, to support allies, and to begin the complicated processes of positioning forces for combat and preparation of the environment.

1-24. During conflict, ARSOF may be most effective in conducting strategic and operational economy-of-force operations. Under certain circumstances, the SOAR may generate military or diplomatic advantages disproportionate to the resources they represent. With support from the SOAR, SOF can locate, seize, or destroy strategic and operational targets and obtain critical intelligence. SOF can disorganize, disrupt, and demoralize an enemy commander's troops and divert important enemy resources. SOAR attack aircraft can act unilaterally to destroy enemy forces and equipment with precision fires.

1-25. ARSOF expands the availability of options of the President of the United States and the Secretary of Defense, particularly in crises and contingencies that include counterterrorism, insurgency, subversion, and sabotage. These contingencies normally are diplomatic initiatives where the overt uses of large conventional forces are impractical. The use of ARSOF enables decisionmakers to prevent a conflict or to limit its scope. Decisionmakers can therefore better control committed U.S. forces and resources. The SOAR in support of SOF may be the best choice for actions requiring a rapid response or a surgically precise and focused use of force.

1-26. The SOAR habitually trains in small groups or as part of an integrated U.S. response with other military forces, as well as non-Department of Defense and civilian agencies. Selected small, self-contained units can work swiftly and quietly without the noticeable presence of conventional forces. Even under the most austere conditions, these units are able to operate without the infrastructure often needed by a large force.

1-27. To ensure missions selected for the SOAR are compatible with ARSOF capabilities, commanders must be familiar with the following SO fundamentals and characteristics:

- SOAR personnel undergo a careful selection process and mission-specific training beyond basic military skills to achieve entry-level SO skills. These selection and training programs make it unlikely that rapid replacement or reconstitution of personnel, capabilities, and equipment can easily be accomplished.
- Mature, experienced personnel compose the SOAR. Many maintain a high level of competency in more than one military specialty.
- Some elements are regionally oriented for expedient employment. Cross-cultural communications training and education are routine parts of the regional orientation.
- The SOAR conducts specific tactical operations by small units with unique talents that directly strike or engage strategic and operational aims and objectives.
- Planning for SO may begin at the unified, joint, or interagency levels for execution that requires extensive and rigorous rehearsal.
- SO are frequently clandestine or low-visibility operations. Occasionally, SO may be a part of an overt operation. They can be covert but, as such, require a declaration of war or a specific finding executed by the President and/or the Secretary of Defense. ARSOF can deploy at a relatively low cost with a low profile that is less intrusive than that of larger conventional forces.
- SO units often conduct missions at great distance from their operational bases. These units employ sophisticated communications systems that support insertion, sustainment, and extraction from hostile, denied, or politically sensitive areas.
- SO occur throughout the range of military operations, to include military engagement, security cooperation and deterrence, crisis response and limited contingency operations, and major operations and campaigns.

- SO strive to influence the will and decisionmaking process of foreign leadership that create conditions favorable to U.S. strategic aims and objectives.
- SOAR missions are often high-risk operations that have limited windows of execution and require first-time success.
- Employment of SO may require patient, long-term commitment and support to achieve U.S. national goals in an operational area.
- The SOAR requires theater and, frequently, national-level intelligence support.
- Selected SO require a detailed knowledge of the cultural nuances, social mores, and languages of a country or region where employed.
- The SOAR's mission is inherently joint and at times multinational, which requires interagency and international coordination. The contribution and success of ARSOF to national objectives is greatest when ARSOF are fully integrated into the joint force commander's plan at the earliest stages of planning.
- The SOAR can be task-organized quickly and deployed rapidly to provide tailored responses to many different situations and contingencies.
- The SOAR can gain access to hostile and denied areas.
- The SOAR can live and operate in austere, harsh environments without extensive support for short durations. For long-duration operations, the SOAR requires support from the Army Service component command.
- ARSOF can survey and assess local situations and report these assessments rapidly.

SOAR RESPONSIBILITIES IN SUPPORT OF ARSOF CORE ACTIVITIES

1-28. There are driving principles of the SOAR's responsibilities in support of ARSOF. These principles are as follows:

- Infiltrate, sustain, and exfiltrate U.S. SOF and other selected personnel.
- Insert and extract SOF land and maritime assault vehicles and vessels.
- Conduct direct action, close combat attack, and close air support operations using organic attack helicopters to provide aerial firepower and terminal guidance for precision munitions, unilaterally or with other SOF elements.
- Provide forward air control for U.S. close combat attack, multinational close air support, and indirect fires.
- Conduct special reconnaissance missions.
- Conduct intelligence, surveillance, reconnaissance, and target acquisition.
- Conduct limited electronic warfare.
- Recover personnel or sensitive materiel.
- Conduct assisted evasion and recovery when dedicated combat search and rescue assets are unavailable.
- Conduct combat search and rescue as a part of the SOF component apportioned to the Joint Personnel Recovery Center when the mission requires capabilities above and beyond conventional theater combat search and rescue assets.
- Perform emergency air evacuation of SOF personnel during the conduct of SO.
- Conduct strategic self-deployment of all aerial refuel-capable helicopters.
- Conduct SO joint maritime operations.
- Conduct SO water insertion and recovery operations.
- Support and facilitate ground and aerial C2, communication and computer systems, and reconnaissance and intelligence operations.
- Provide the C2 element for SOA assets and attached conventional aviation assets supporting SOF.

- Accept augmentation, as needed, from United States Air Force assets, military occupational specialties (MOSs), and equipment.
- Provide liaison officers or subject-matter experts as required, subject to availability.
- Perform aviation unit maintenance and aviation intermediate maintenance for all organic aircraft.

PRINCIPLES OF SOAR EMPLOYMENT

1-29. SO principles are an important part of SO mission planning. Principles for the SOAR are as follows:
- Integrate supporting SOA assets from mission analysis to course of action development through mission accomplishment.
- Increase SOAR effectiveness by using the tactical and logistic capabilities of other Services and nations.
- Use near-real-time and all-source intelligence products during mission planning, rehearsal, and execution.
- Negate hostile acquisition means and weapons systems before and during the mission.
- Employ the element of surprise by—
  - Conducting operations at night and during periods of low ambient light.
  - Using deception and operations security measures.
  - Using terrain-following techniques.
  - Using the range capability of the aircraft to fly indirect approaches.
  - Controlling or reducing electronic emissions during the mission.

1-30. During extended operations, the SOAR must—
- Change tactics and procedures regularly to avoid becoming predictable.
- Anticipate enemy actions.
- Concentrate combat power on enemy vulnerabilities.
- Attempt to continue to shorten the decision cycle to mission execution.
- Stay flexible.
- Clearly designate and articulate the main effort.
- Move SOF throughout the depth of the operational area as the tactical situation changes.
- Concentrate SOF at the critical time by using precision timing and navigation.
- Displace forward elements frequently for security.
- Maintain the ability to operate continuously.
- Understand the effects of battle on Soldiers, units, and leaders.

EMPLOYMENT CONSIDERATIONS

1-31. The SOAR provides SOF the capability to penetrate hostile or denied territory. To accomplish SO core activities, SOAR units have specialized aircraft with sophisticated state-of-the-art special mission equipment. SOAR aircrews undergo intense training in the tactical employment of the aircraft and the execution of SO aviation responsibilities. The SOAR normally arrives in the theater of operations with other SOF before hostilities begin. Because of the SOAR's training and equipment, they are normally deployed against high-payoff targets that support the joint force commander's campaign plan. The SOAR exploits the darkness, adverse weather conditions, and extended range and navigation systems of their aircraft to penetrate hostile territory from unexpected avenues of approach in the execution of SO missions.

ATTACK HELICOPTERS

1-32. The accuracy and lethality of attack helicopters make them useful in supporting egress or ingress when conducting operations. The SOAR possesses attack helicopter capability that is adequate for ARSOF missions. "Close air support" is a joint term, and the Army does not consider its helicopters close air

Chapter 1

support systems. The SOAR can conduct attacks employing close air support (joint tactics, techniques, and procedures) when operating in support of SOF and other forces, and a Level 1 close air support controller is in control of weapons release. Further, SOAR can conduct deep attack against enemy objectives and equipment as required by the supported commander. Joint Publication (JP) 3-09.3, *Close Air Support*, provides further guidance on close air support.

1-33. Close combat attack in support of SOF or other forces does not require the presence of a Level 1 close air support Service member. Identification of the target is coordinated between the designated ground Service member and the air mission commander. The air mission commander is responsible for developing the situational awareness and understanding of friendly and enemy locations, ensuring positive identification of the target and synchronization for weapons release. Detailed information on close combat attack is in FM 3-04.126, *Attack Reconnaissance Helicopter Operations*.

1-34. The limited availability of attack helicopters, as a theater asset, and their versatility place them in great demand. In situations where the SOAR's attack helicopter assets are inadequate to support a directed SOF mission, the SOAR commander must request additional assets as soon as possible to ensure adequate planning and training. When making such requests, the commander must state his intent and plan for the use of the additional helicopters.

1-35. When planning operations, staging the SOAR assets from the same base allows face-to-face briefings, which in turn improves mission coordination. Staging from the same base may also reduce or mask the operational visibility of the SOAR's operations. However, there may be operational conditions where the grouping of all assets at the same location may adversely impact operational security. The commander must evaluate all security and operational factors when deciding how to base the SOAR assets.

RECONNAISSANCE

1-36. The ARSOF core activity of special reconnaissance is discussed on page 1-3 of this manual. Immediate, unplanned, or uncoordinated reconnaissance by air assets is normally unavailable. When requested, the procedures for requesting the support are the same as those for close air support or close combat attack.

SECURITY

1-37. Security concerns directly impact SO planning and mission execution. However, excessive compartmentalization can also prove detrimental to mission success by excluding key personnel from the planning process. SOAR commanders must resolve these conflicting demands on mission planning and execution. Although insufficient security can compromise a mission, excessive security can jeopardize the coordination of plans, which can contribute to loss of life and critical equipment that may lead to mission failure.

1-38. Compartmented planning is typical for SO and planning staffs are normally small. Within a compartmented activity, however, individuals must share information.

1-39. To enhance security and achieve surprise, individuals must follow intelligence, counterintelligence, electronic warfare, and cover and deception procedures when planning and executing SO. To provide security for the plan and to preserve the security for other planned operations, ground and air planners must go into isolation.

1-40. To enhance operational security, ground forces normally discuss only the portion of the ground tactical plan that involves the SOAR crews—the immediate actions on the objective and actions in the case of loss of aircraft. Planners must coordinate and deconflict air operations, to include discussion of evasion and recovery plans and combat search and rescue procedures for downed aircrews. This practice provides mutual protection for all SOF personnel and the mission in the event of capture or compromise.

MANEUVER

1-41. Maneuver seeks to place the enemy in a position of disadvantage using the flexible application of combat power. During SO, maneuver implies the ability to infiltrate and exfiltrate hostile areas by exploiting enemy weaknesses. Successful maneuver gives the SOAR the ability to infiltrate the supported commander's elements, to strike the enemy where and when the enemy is most vulnerable, and to successfully exfiltrate friendly forces by avoiding the enemy's reaction forces and strengths.

1-42. SOAR tactics focus on surprise, economy of force, maneuver, and simplicity. Mission execution should be during the hours of darkness, as risks increase during daylight operations. Low-level terrain flight altitudes during low ambient light or limited visibility provide the element of surprise. Training in night formation flight and precision navigation enables the massing of combat power at the precise time and place. The use of indirect routes exploits the increased range capability of the aircraft and is a measure to avoid known enemy locations or indigenous personnel. This capability allows the SOAR to maneuver over the operational environment. Simplicity is only possible because of the high level of training of the SOAR's aircrews and the equipment employed by the SOAR. All SOAR aircraft are capable of precision navigation, long-range secure communications, long-range flight performance, and increased weapons lethality.

COMMAND AND CONTROL

1-43. Unless otherwise directed by the Secretary of Defense, all SOF based in the United States are under the combatant command of the Commander, USSOCOM. SOF assigned to a theater are under the command of the geographic combatant commander. Within a geographic combatant command, C2 of SOF should be executed within the SOF chain of command. Normally, C2 of SOF is executed through the TSOC or a JSOTF. To fully integrate SO and conventional operations, SOF must maintain effective liaison with all components of the joint force to ensure that unity of effort is maintained and risk of fratricide is minimized.

1-44. There are several options for C2 of the SOAR. These options are dependent upon the mission and the task force organization structure that has been determined best to achieve mission success. JP 3-05, *Special Operations*, and FM 3-05, *Army Special Operations Forces*, include detailed information on the C2 of the SOAR and ARSOF.

1-45. The JFSOCC normally delegates C2 of the SOAR to the JSOACC or gives operational control to the SOF ground commander. The JSOTF commander is the theater JFSOCC. In stability operations, the JSOTF commander reports directly to the geographic combatant commander. In war, he reports directly to the joint task force commander. The JSOTF commander is the principal SO advisor in-theater. All SOF normally fall under his control. The JSOTF commander may organize the JSOTF headquarters (HQ) as necessary to carry out all assigned duties and responsibilities. The commander may retain command of the SOAR task force on an equal level with the conventional force commanders.

1-46. Dependent upon mission requirements, it may be determined that the JSOACC may not be the best method to C2 the SOAR. Experience has shown that in certain circumstances efficiencies are gained when SOAR assets are directly under the C2 of the combined JSOTF, who task-organizes the SOAR to best meet mission requirements.

This page intentionally left blank.

Chapter 2
Mission Command

THE JOINT OPERATION PLANNING AND MILITARY DECISIONMAKING PROCESSES

2-1. The joint operation planning process and the military decisionmaking process (MDMP) are planning models that establish procedures for analyzing a mission. They assist the commander in developing, analyzing, and comparing courses of action against one another. This chapter primarily discusses the MDMP. JP 5-0, *Joint Operation Planning*, includes further information on the joint operation planning process and links to national strategic planning.

2-2. The MDMP assists the commander and his staff in selection of the optimum course of action, and in producing a plan or order. The MDMP applies across the spectrum of conflict and range of military operations. Commanders and staffs use the MDMP to organize their planning activities, share a common understanding of the mission and commander's intent, and to develop effective plans and orders. Appendix A contains the formats for a SOAR operation order and the mission planning folder.

2-3. The MDMP helps organize the thought process of commanders and staffs. It helps them apply thoroughness, clarity, sound judgment, logic, and professional knowledge to reach decisions. The shaded boxes in Figure 2-1, page 2-2, depict the seven steps of the MDMP. Each step begins with inputs that build upon previous steps. The outputs of each step drive subsequent steps. Errors committed early in the planning process affect later steps and lead to accumulated errors that may lead to mission failure. Although the formal process begins with the receipt of a mission and has as its goal the production of a warning order, planning continues throughout the operations process.

2-4. The products created during the full MDMP are for use during subsequent planning sessions when time may be unavailable for thorough planning, and existing mission variables of mission, enemy, terrain and weather, troops and support available, time available, and civil considerations are substantially unchanged. The desired outcome of an effective mission-planning process is the synchronization of total combat power in the operational environment.

2-5. Preparation and execution, although not part of the MDMP, are shown in the lower portion of Figure 2-1 to highlight the importance and necessity of continuous planning and dialogue. FM 5-0, *The Operations Process*, includes more detailed information on the MDMP.

PARTICIPATION DURING THE JOINT OPERATION PLANNING AND MILITARY DECISIONMAKING PROCESSES

2-6. The SOAR commander, his staff, and subordinate commanders and flight leads must be incorporated into the MDMP or joint operation planning process at the earliest opportunity. Clear identification and understanding of the C2 structure in which the SOAR is going to operate is critical to mission success. The SOAR normally interacts with the JSOTF, the joint special operations air component (JSOAC), the special operations liaison element (SOLE), and the Army and Navy ground force elements.

JOINT PLANNING CONSIDERATIONS

2-7. SOAR operations are inherently joint in nature. The nature of modern warfare demands ARSOF plan and fight as a team with its sister Services. For joint forces to win in battle, they must have a single, unified planning and execution framework capable of translating individual Service terminology and operational policies into a commonly understood language and standing operating procedures.

Chapter 2

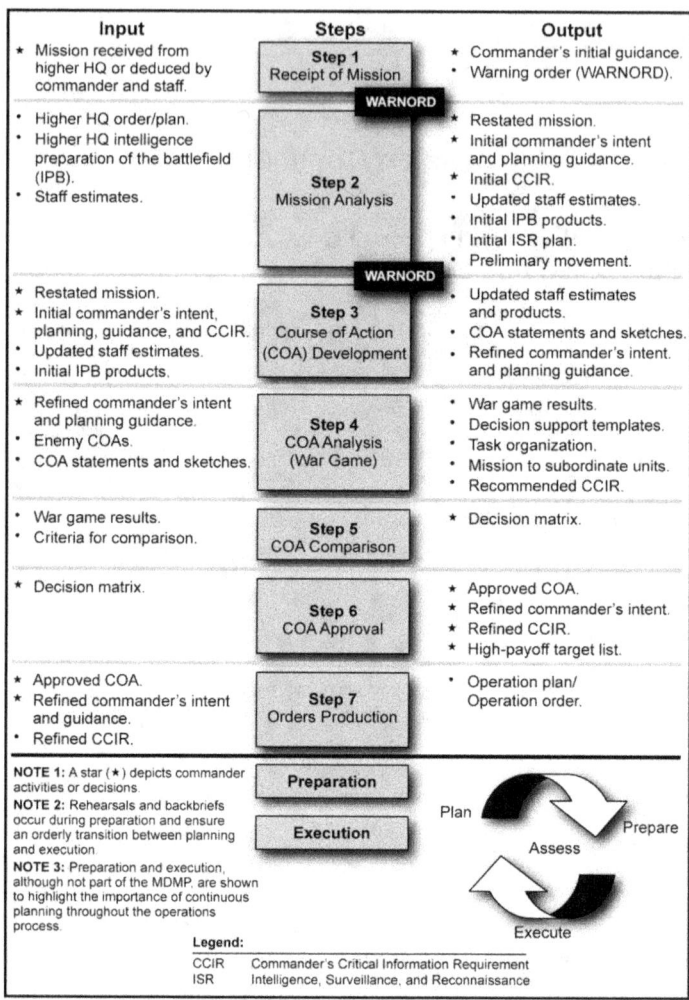

Figure 2-1. The military decisionmaking process

2-8. The joint operation planning process combines individual Service terminology and operating procedures into one standard multifaceted system. It provides standardization to the joint planning system used for the execution of complex multi-Service exercises, campaigns, and major operations. The joint operation planning process uses a set of C2 techniques and processes, supported by a computerized information system, to ensure the right amount of timely support gets to the warfighter to ensure a decisive victory. The principles of war and joint operations that guide the joint operation planning process are depicted in Figure 2-2, page 2-3. JP 5-0 contains further information on the joint operation planning process.

Mission Command

Figure 2-2. Principles of war and joint operations

JOINT SPECIAL OPERATIONS AIR COMPONENT COMMANDER

2-9. The SOAR may operate under the operational control of a JSOACC designated by the JFSOCC (Figure 2-3). The SOAR may also operate in direct support to the ground commander or JSOTF/JFSOCC with a coordination line to the JSOAC for integration into the overall air plan (Figure 2-4, page 2-4). The JSOACC is the Service SO air commander who has the majority of SOA forces and is most capable of providing C2. The JSOACC deconflicts and coordinates SOA with conventional air operations by direct coordination with the joint force air component commander. If more than one aviation unit or Service is present, a JSOACC unifies the C2 of aviation assets under a single air manager. The JSOACC provides the command the most efficient use of aviation assets to meet mission requirements. With proper personnel and equipment augmentation, the SOAR or battalion commander and his staff could also serve as a JSOACC.

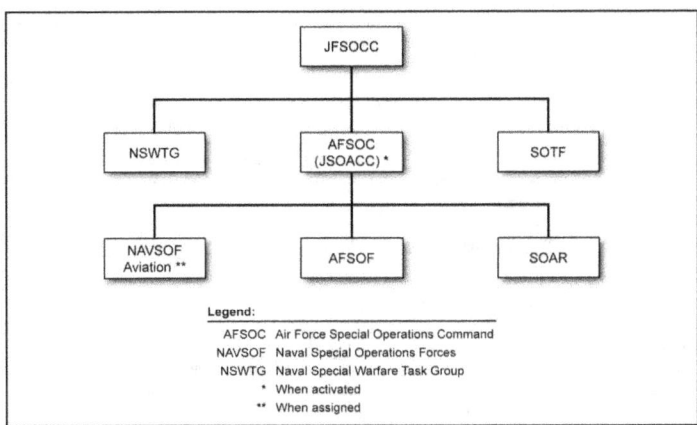

Figure 2-3. Typical JSOACC command relationships

Chapter 2

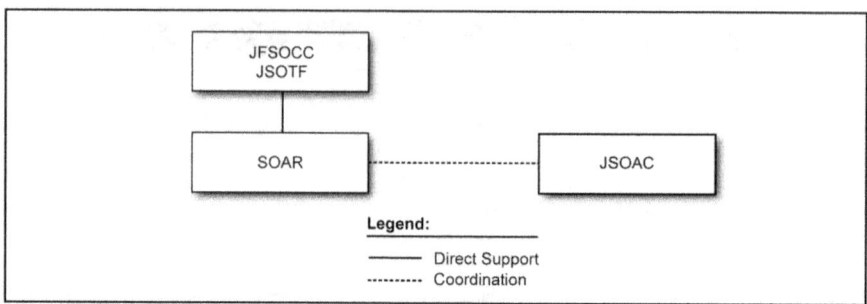

Figure 2-4. Direct support relationship

SPECIAL OPERATIONS LIAISON ELEMENT

2-10. The SOLE (Figure 2-5) provides SOF liaison to the joint force air component commander and reports directly to the JSOTF. The SOLE is a joint SOF organization that includes SOF aircrews, intelligence, airspace, logistics, and special tactics teams, to include combat control and pararescue teams, Army SF, and Navy sea-air-land teams (SEALs). It ensures that all SOF targets, teams, and air missions are deconflicted, properly integrated, and coordinated at all planning and execution phases. Specific functions include air tasking order and airspace control order development, real-time mission support within the joint force air component command, operations and intelligence support for targeting, combat airspace control for prevention of fratricide, coordination with special plans and functions, and coordination with the joint search and rescue center. The SOLE coordinates and synchronizes SOF air and surface operations with conventional air operations. The SOLE must consider airborne fire support and reconnaissance C2 aircraft, aerial refueling, and deconfliction of deep-battlefield operations. The SOLE also assists the joint force air component commander in deconfliction of the JSOA. The SOAR primarily interacts with the SOLE through the JSOAC, which ensures that SOAR operations are deconflicted and supported with unconventional assets, as required.

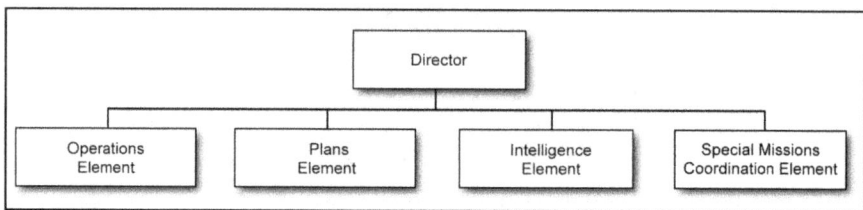

Figure 2-5. Typical special operations liaison element organization

AIRCRAFT CONSIDERATIONS

2-11. The SOAR has three basic types of rotary-wing aircraft: light assault, medium lift/defensive armed penetrator (DAP), and heavy lift. Each SOAR aircraft has undergone significant modification and upgrades that have greatly increased the capabilities and survivability of the aircraft and its crews. Appendix B includes a detailed description and capability for each aircraft.

AIRSPACE COMMAND AND CONTROL

2-12. An important aviation link in the planning and execution process is airspace coordination and deconfliction. This responsibility is handled by the JSOACC and the joint force air component commander through the SOLE. He deconflicts JSOTF fixed- and rotary-wing missions to ensure effective use of the forces and to prevent fratricide. The joint force air component commander is normally located at the joint force command. He deconflicts theater assets and produces allocation requests, air tasking orders, and airspace control orders.

2-13. The C2 of airspace in the combat zone enhances combat operations by promoting safe, efficient, and flexible airspace use. The SOAR must be aware of positive procedural Army airspace C2 measures. In addition, the SOAR must adhere to directed control procedures. To facilitate rapid and accurate dissemination of information on the use of airspace, the commander must make sure a strong, ongoing communications link exists with Army airspace C2 elements.

AIRSPACE DECONFLICTION

2-14. Route deconfliction is vital to SOAR survival. Review of routes must, therefore, occur at several levels. Planners should deconflict their routes by plotting them on a map that is current with the latest airspace control measures and enemy locations. They should also send the route information through the chain of command to the Army airspace C2 cells for input into the airspace deconfliction software.

2-15. When deconflicting the route of flight, planners should consider time and space, and plot all routes of flight on the map to make sure the routes do not conflict with any airspace control measure published in the airspace control order. They should then note all locations where routes cross and make sure aircraft will not cross within 10 minutes of each other. To allow for changes in mission times, all mission aircrews should be notified of the aircraft crossing routes within 30 minutes of each other.

2-16. Mission planners must send the route information up the chain of command as soon as the information is available. The information for the air tasking order must be submitted not later than 24 hours before the air tasking order takes effect. The lead time is necessary so that aerial-refueling requests can be submitted and put into the air tasking order.

2-17. In undeveloped areas of operation, or during initial stabilization, airspace control plans should be established throughout the area of operations. These airspace control plans expedite the planning process and enable Army airspace C2 personnel to plot and deconflict the routes more quickly and accurately. Mission planners and the airspace control authority should develop the list of airspace control plans. If an airspace control order has been published, those airspace control plans should be used.

AIRSPACE CONTROL MEASURES

2-18. There are many airspace control measures that provide procedural control. These include the following:
- *High-density airspace control zone*: A high-density airspace control zone is an area of concentrated employment of numerous and varied weapons or airspace users. The zone has defined dimensions that usually coincide with geographical features or navigational aids. The appropriate commander normally approves access to a high-density airspace control zone and air defense weapons status within the zone.
- *Coordinating altitude*: A coordinating altitude is a procedural method to separate fixed- and rotary-wing aircraft by determining an altitude below which fixed-wing aircraft normally will not fly and above which rotary-wing aircraft normally will not fly. It may include a buffer zone for small altitude deviations and extend from the forward edge of the communications zone to locations forward of the recognized line of troops. The coordinating altitude does not restrict fixed- or rotary-wing aircraft when operating against or in the immediate vicinity of enemy ground forces. Fixed- or rotary-wing aircraft planning an extended penetration of this altitude should notify the appropriate airspace control facility. Approval acknowledgment is not, however, required before fixed-wing aircraft operate below the coordinating altitude or rotary-wing aircraft operate above the coordinating altitude.
- *Restricted operations zone*: The terms "airspace restricted area" and "restricted operations zone" refer to the same control measure. A restricted operations zone is a volume of airspace of

defined dimensions developed for a specific mission. Some or all airspace users are restricted from the area until the end of the mission. A restricted operations zone may be established around a tactical airfield, drop zone, search and rescue operation, infiltration and exfiltration points, and special electronic mission aircraft orbits, or it may be established to facilitate fire support operations.
- *Joint special operations area*: The JSOA is a restricted area of land, sea, and airspace assigned by a joint force commander to the commander of joint SOF to conduct SO activities. The commander of joint SOF may further assign a specific area or sector within the JSOA to a subordinate commander for mission execution.
- *Low-level transit route*: A low-level transit route is a temporary corridor of defined dimensions in the forward area. It minimizes the risk to friendly aircraft from friendly air defenses or surface forces.
- *Minimum-risk route*: A minimum-risk route is a temporary corridor of defined dimensions recommended for use by high-speed, fixed-wing aircraft. It presents the minimum known hazards to low-flying aircraft transiting the combat zone.
- *Standard Army aircraft flight route*: A standard Army aircraft flight route is a route below the coordinating altitude established to facilitate the movement of Army aircraft during visual meteorological conditions. Army aircraft movements in an unassigned operations area during instrument meteorological conditions will comply with established instrument flight rules. The standard Army aircraft flight route is normally in the corps and the division unassigned areas. Unassigned areas refer to portions of an area of operations not assigned to a subordinate organization. Unassigned areas remain the responsibility of the HQ responsible for the entire area of operations. The Army airspace C2 element develops standard Army aircraft flight routes to route Army aircraft safely when conducting sustainment and combat support of missions at terrain flight altitudes. Standard Army aircraft flight routes are primarily for single aircraft or for small flights of aircraft.
- *Base defense zone*: A base defense zone is an air defense zone established around an air base. It is limited to the engagement envelope of the short-range air defense systems defending the base. Base defense zones have specific entry, exit, and identification, friend or foe, procedures aircrews must follow.
- *Weapons-free zone*: A weapons-free zone is an air defense zone established for the protection of key assets or facilities of the joint force other than air bases. Air defense artillery systems defending the weapons-free zone may fire at any target not positively identified as friendly. Aircrews must avoid active weapons-free zones or coordinate their use with the designated control authority before entering or transiting the zone.
- *Air corridor*: An air corridor is a restricted air route of travel specified for friendly aircraft use to prevent fires against friendly aircraft. Air corridors are temporary corridors for routing combat elements of the division and corps aviation brigade between such areas as assembly areas, battle positions, and forward arming and refueling points. Air corridors are control measures employed during air assault operations to designate routes for air assault forces during the air movement phase.
- *Underdeveloped countries*: Although the SOAR often operates in underdeveloped countries, SOAR personnel should review established airspace control measures. The control measures help planners avoid suspected concentrations of aircraft. When the operations are in developed countries, the airspace control measures show concentrations of aircraft as well as other hazards to flight.

PLANNING PROCESSES

2-19. The SOAR and the JSOTF use the SO mission-planning process depicted in Figure 2-6, page 2-7. This process complements the C2 structure. The timeline is a base timeline for mission support, but the six mission variables can increase or reduce the timeline as required. The timeline generates from the earliest anticipated launch time, which is the JFSOCC's best estimate for mission execution. The JSOAC must keep the JSOTF informed of the SOAR's asset availability, as limited aviation platforms may affect the

JSOTF's operating tempo and hinder mission execution. The JFSOCC sends the tasking order to the joint air operations center and simultaneously to the ground force element and the SOAR.

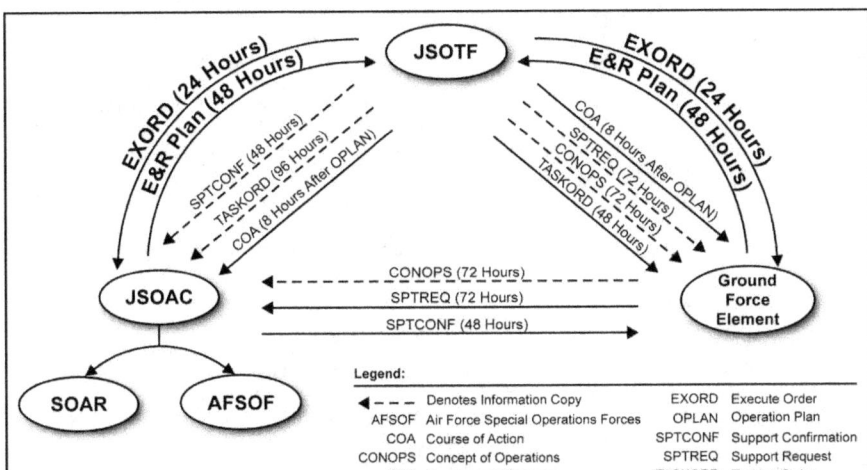

Figure 2-6. The special operations mission-planning process

2-20. The mission flow defines the JSOTF planning sequence and provides limits and boundaries to the mission process. The SOAR integrates the supported SOF ground commander's plan into the mission-planning process. The SOAR representative to this process is the aviation liaison officer attached to the planning staff. The joint force special operations component (JFSOC) representative to the process is the SOLE. The JFSOC requires the SOLE to coordinate, deconflict, and integrate SOF air and surface operations with conventional air operations. The SOAR receives the information copy of the tasking order at approximately the same time as the planning staff receives the support request.

2-21. The SOAR conducts parallel planning, allowing the liaison officer to receive initial guidance from the aviation commander and to input those limitations and constraints into the planning staff's course of action development. Liaison officer and SOLE inputs at this stage are critical in ensuring the feasibility of aviation survivability and support for the ground force commander's course of action. This early input reduces planning time through elimination of impracticable aviation courses of action. The liaison officer participates in the war games with the planning staff to determine decision points and abort criteria critical to mission success. The SOLE ensures all support criteria for the mission are provided to the joint force air component commander. Throughout the entire process, the liaison officer and SOLE keep higher HQ and the SOAR operations officer informed of the mission's direction and the commander's critical information requirements and intent.

2-22. The liaison officer participates in the operations order brief to the supported ground force commander. The ground force commander coordinates with the liaison officer to brief the preliminary ground plan and rehearsal plan. This first meeting with the ground force element is usually 8 to 12 hours after the ground force commander's mission brief. During this period, the ground force element conducts its mission planning and completes a detailed aviation mission checklist. The checklist discusses infiltration, exfiltration, contingencies, communications, and other requirements.

2-23. The liaison officer takes the preliminary tactical plan, the rehearsal plan, and the completed checklist to the aviation operations officer for analysis. The liaison officer then disseminates the information to the air mission commander and the flight leader. If the rehearsal is complicated, the flight leader takes that responsibility from the liaison officer and conducts the rehearsal with the ground force element to develop the plan. The next meeting is not later than the ground force element's backbrief to the ground force commander. The flight leader finalizes the tactical plan, the evasion and recovery plan, the communications plan, and any other contingencies that may need adjustment after the rehearsals. The liaison officer and flight leader disseminate threat and mission updates from the ground force element until mission launch. The air mission commander, the flight leader, and the ground force commander perform simulations to examine the threat and to develop appropriate contingencies for flying given routes, using the special operations forces planning and rehearsal system (SOFPARS). The planning and rehearsal information support provides information to SOFPARS during rehearsal and simulation.

2-24. The SOAR performs a tailored MDMP that parallels the ground force commander's planning process. Several factors favor the MDMP. The primary SOAR mission is to nurture the SOAR's relationship with the ground force and to support the force with as many assets as the mission requires. This type of support forces the SOAR to react and adjust to the ground commander's tactical plan, thus limiting the SOAR's courses of action and planning time.

2-25. Throughout mission planning, the SOAR must remain flexible and adaptable to the ground force commander's intent. Without the ground plan, the SOAR's courses of action are limited to asset availability, forward arming and refueling point capability, and scheme of maneuver. The SOAR can develop and request suppression of enemy air defenses and fire support through the SOLE, but the ground force commander must integrate those requests into the tactical plan. The SOAR war-games the entire tactical plan and finalizes full-mission profile rehearsals only after the ground force commander approves the ground force element's plan.

2-26. Special reconnaissance and FID missions may not require complex analysis and may require only static rehearsals, such as rock and contingency drills, because these missions usually require only infiltration and exfiltration operations. Direct action missions are, however, normally intricate operations that require detailed war-gaming and flying rehearsals. Synchronizing the planning, focusing the key players, conducting rehearsals, and performing precombat checks and inspections are critical to the SOAR mission-planning process and to mission success.

2-27. Figures 2-7 through 2-9, pages 2-9 and 2-10, illustrate the sequence of Army SOA mission planning, including the tasks performed by each mission element. The mission-planning timeline begins 106 hours before takeoff and progresses to mission takeoff, termed "H-hour."

PREVENTION OF FRATRICIDE

2-28. Fratricide is defined as the unintentional killing or wounding of friendly personnel by friendly fire. Although occasionally the result of malfunctioning weapons, fratricide has historically been the result of human error and confusion on the battlefield.

2-29. The destructive power and range of modern weapons, coupled with the high intensity and rapid tempo of modern combat, increase the potential for fratricide. Commanders must be aware of those situations that increase the risk of fratricide and, in response, institute appropriate preventative measures.

2-30. The primary mechanisms for reducing fratricide are command emphasis, disciplined operations, close coordination among component commands, rehearsals, and enhanced situational awareness. These are especially true for ARSOF since they are frequently deployed forward of the fire support coordination line. Commanders must make every effort to reduce the potential for fratricide while not limiting boldness and initiative of the forces in combat.

Mission Command

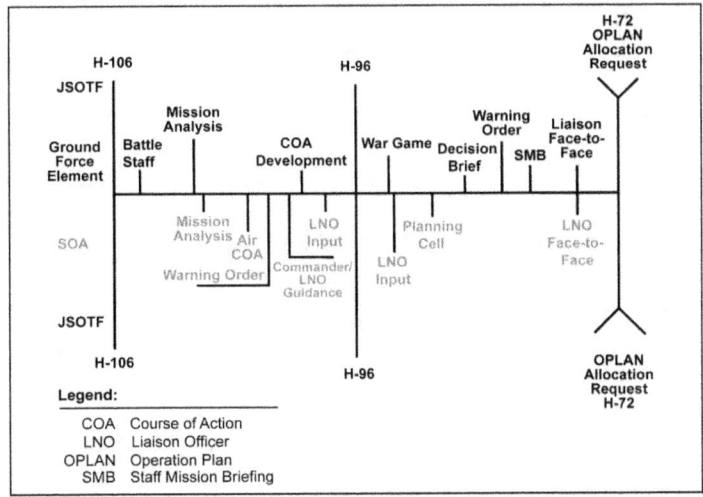

Figure 2-7. Timeline phase, 106 hours to 72 hours

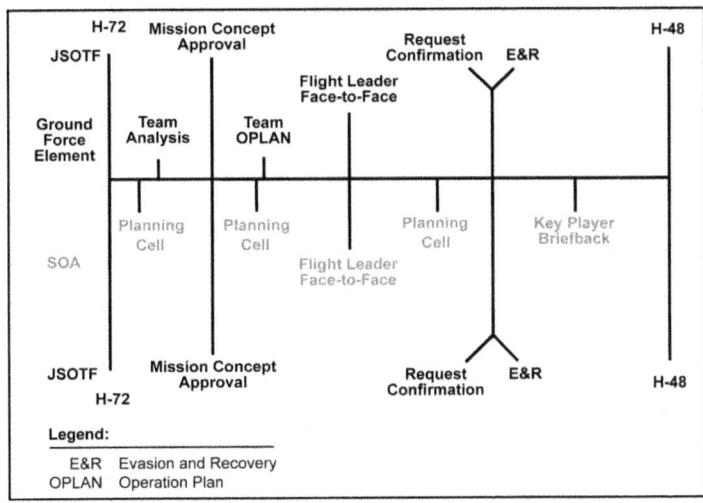

Figure 2-8. Timeline phase, 72 hours to 48 hours

28 October 2011 FM 3-76 2-9

Chapter 2

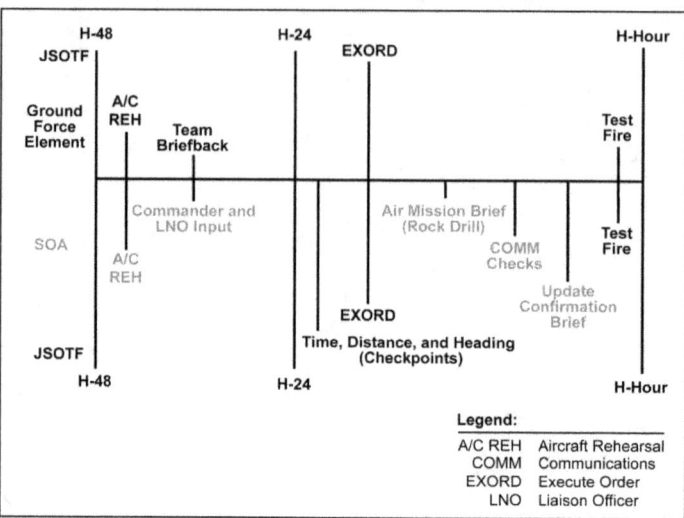

Figure 2-9. Timeline phase, 48 hours to takeoff (H-hour)

Chapter 3
Operations

FORCE DEPLOYMENT, EMPLOYMENT, AND REDEPLOYMENT

3-1. The SOAR is prepared on short notice to support SOF across the full range of military operations. These SOF elements may be in support of other U.S. military forces, host nations, or other U.S. agencies and special activities. To support directed requirements successfully, the SOAR maintains a high level of readiness, exploited by continuous training in all operational environments.

PREDEPLOYMENT

3-2. SOAR predeployment consists of activities conducted by the unit before the execution of the mission to improve its ability to achieve mission success. Such activities may include, but are not limited to, the concept plan refinement, rehearsals, reconnaissance, coordination, inspections, standing operating procedure reviews, load plan verification, Soldier readiness preparation, weapons test-firing, and movement. The complexity of the operation may impose significant challenges. The characteristics of land and sea operations differ tremendously from situation to situation. Therefore, it is imperative that the SOAR maintain proficiency in all environments, to include desert, mountain, maritime, urban, arctic, and jungle. Mission success depends upon honing basic and advanced skills and exercising the mission-essential task list as much as effective and detailed planning. Rehearsals help staffs, units, and individuals prepare for execution of the full spectrum operations.

Staff Preparation

3-3. Each staff section and element conducts activities to maximize the operational effectiveness of the regiment and SOTF. Coordination between echelons and mission preparation that precedes execution are just as important as developing an effective plan. Staff preparation includes assembling and continuously updating estimates. For example, continuous intelligence preparation of the operational environment provides accurate situational updates for commanders. Updated estimates form the basis for staff recommendations; the value of current, reasonably accurate estimates increases exponentially with tempo. Whether incorporated into a formal process or not, the preparatory activities of staff sections and SOAR force elements ensure updated planning and operational refinement is continued throughout the preparation and execution phases of the full spectrum of military operations.

Unit Preparation

3-4. Failure to train to standards exposes the SOAR to additional risks that are incompatible with mission success. Warrior skills should be continuously developed and honed. The SOAR habitually conducts realistic training with organizations it routinely supports. The SOAR's continuous ability to train and educate its forces increases its ability to fight and win. Without a rigorous commitment to realistic training, deployment of the SOAR to a theater may very well entail unacceptable risks.

3-5. Preparation and responsiveness also emphasize training, planning, and preparation for deployment. Commanders recognize that crises rarely allow sufficient time to correct training deficiencies between alert and deployment. Commanders ensure that their units are prepared to accomplish their mission-essential tasks before alert and to concentrate on mission-specific training in the time available afterward. Today's operational environment demands that the SOAR remains continuously combat-ready and able to quickly adapt to noncombat situations or stability operations. SOAR missions by their nature are normally high-risk.

Chapter 3

DEPLOYMENT PLANNING

3-6. Deployment planning is planning to move SOAR forces and their sustainment resources from their original locations to a specific operational area to conduct joint operations. Deployment planning is the responsibility of the combatant command's Army Service component command in close coordination with the 528th Sustainment Brigade (Special Operations) (Airborne) (SB[SO][A]) and the United States Transportation Command. It involves planning for the continental United States (CONUS), intertheater (strategic), and intratheater movement of forces and the required resources to sustain them. Strategic-deployment planning focuses on the intertheater movement of forces and resources, using national, allied, and coalition strategic-deployment capabilities. Due to the low density and high demand of SOAR assets, deployments should be conducted via intertheater airlift to minimize lost asset time due to movement to/from an operation.

3-7. Deployment is the movement of forces and materiel from their point of origin to the theater. This process has four supporting components—predeployment, fort-to-port, port-to-port, and port-to-destination (reception, staging, onward movement, and integration) activities. More information on the deployment process is contained in JP 3-35, *Deployment and Redeployment Operations*, and FM 3-05.140, *Army Special Operations Forces Logistics*.

3-8. Commanders continuously plan, prepare, execute, and assess operations with initial-entry forces, while assembling and preparing follow-on forces. To seize the initiative during deployment and the early phases of an operation, commanders accept calculated risks, even when the enemy situation is not well developed. Balancing these dynamics is an art mastered through study and experience.

EMPLOYMENT PLANNING

3-9. Employment planning prescribes how to apply force to attain specified military objectives. Employment planning concepts are developed by the combatant commanders through their component commands. Employment planning provides the foundation and determines the scope of mobilization, deployment, sustainment, and redeployment planning. Employment planning considerations that directly impact deployment operations include—
- Identification of force requirements.
- Commander's intent for deployment.
- Time-phasing of personnel, equipment, and materiel to support the mission.
- Closure of the forces required to execute decisive operations.

3-10. Employment planning is discussed in JP 5-0. More information on employment planning is contained in JP 3-0, *Joint Operations*, and FM 3-05.

3-11. Employment is the conduct of operations to support a joint force commander's operations plan. Employment encompasses a wide array of operations, including, but not limited to—
- Entry operations (opposed or unopposed).
- Shaping operations (lethal and nonlethal).
- Decisive operations (combat and sustainment).
- Postconflict operations (prepare for follow-on missions or redeployment).

SUSTAINMENT PLANNING

3-12. Sustainment planning provides and maintains levels of personnel, materiel, and consumables required to sustain the planned combat activity. When planning sustainment, consideration must be given to the duration of the activity at the desired intensity of the operation. FM 3-05.140 includes more information on ARSOF logistics and sustainment.

REDEPLOYMENT

3-13. Redeployment is the transfer of forces and materiel to support another joint force commander's operational requirements, or to return personnel, equipment, and materiel to the home or demobilization station upon completion of the mission. Redeployment operations encompass four phases:
- Recovery, reconstitution, and preredeployment activities.
- Movement to and activities at the port of embarkation.
- Movement to the port of debarkation.
- Movement to the home station (JP 3-35 has detailed information).

3-14. As units start the redeployment phase, the Army Service component command ensures the remaining support units, host nation, or contract personnel are able to meet stay-behind SOAR sustainment requirements.

POSTMISSION ACTIVITIES

3-15. Because many of the SOAR's missions are of short duration, it is incumbent upon the SOAR staff and crews to conduct postmission discussions and activities regarding the overall success of the mission goals, tactics, and equipment performance. Doing so allows corrections to be made to improve the overall performance of the SOAR and other Army aviation units. Postmission activities normally consist of, but are not limited to, the following actions:
- Mission debriefing, to include what was done well and what can be improved upon.
- Resetting of equipment and personnel.
- Publishing of after action review.
- Publishing of observations, insights, and lessons, and forwarding to higher HQ.
- Denoting discrepancies or shortcomings in training and equipment that impacted the mission.

3-16. Further postmission activities may be prescribed in unit standing operating procedures or directed by flight leaders and commanders.

ENVIRONMENTAL CONSIDERATIONS

3-17. The impact of natural and man-made environmental conditions can produce operational risks and benefits. Opportunities for use of terrain, man-made obstacles, darkness, and limited adverse weather conditions to mask or conceal operations must be exploited. However, these same environmental conditions under different circumstances may prove to be a high risk to the overall mission. In such cases, all environmental risks must be evaluated and mitigated.

WEATHER

3-18. Weather affects friendly and enemy capabilities. The SOAR has the unique capability to take advantage of adverse weather conditions and to operate during periods of zero illumination, reduced ceilings, and other limiting meteorological conditions. The effects of the weather must be a part of mission planning. Strong winds affect ground speed. Limited visibility or low cloud ceilings provide some concealment for air operations and aid in achieving surprise. These same conditions, however, restrict supporting high-performance aircraft operations, such as tactical air, close air support, airlift, and aerial refueling. When these conditions exist, close combat attack fire support may still be conducted by rotary-wing assets. Employment of aviation in adverse environments requires in-depth evaluation, study, and knowledge of the geographical terrain to reduce risk and exploit the natural surroundings.

TERRAIN

3-19. Freedom of maneuver over terrain is an inherent characteristic of aviation. This environmental flexibility provides a rapid means to overcome the difficulties of movement and support of ground forces. The use of terrain can be used to an advantage during low-level flight by allowing the aircrew to use vegetation, hills, mountains, and valleys to hide from enemy radar and antiaircraft fire. Several factors must be considered when employing aviation in these environments.

Chapter 3

URBAN ENVIRONMENTS

3-20. In urban environments, aircraft are at high risk from small-arms fire. Urban environments are also complicated by the proximity of noncombatants. Blowing debris increases foreign-object damage and can make landing areas unusable. High levels of artificial illumination reduce night vision goggle perception. If, however, lights are too bright for night vision goggle operation, normal night vision is probably sufficient. Tower and wire hazards increase, thereby causing limited flight routes. Fire control measures must relate to easily distinguishable features to control fratricide. Observation, detection, and weapons engagement of aircraft from numerous locations are critical considerations. Linear corridors, which are characteristic of urban areas, limit fields of fire. Identifying targets, landing zones, and pickup zones can be difficult.

DESERT ENVIRONMENTS

3-21. High daytime temperatures decrease aircraft lift, thus reducing useful loads. Sand and dust increase maintenance requirements and may cause brownouts during takeoffs and landings. Flat, featureless terrain increases the enemy's long-range observation and complicates visual navigation. Sandstorms and other phenomena develop quickly. During desert operations, weather support must be continuously monitored and adverse conditions reported quickly.

MOUNTAIN HIGH-ALTITUDE ENVIRONMENTS

3-22. High altitudes and high temperatures limit lift capabilities, useful loads, and normal cruise speed. Weather conditions change rapidly. Normally, the availability of safe landing areas is significantly reduced. High altitudes place additional demands upon the human body. Because of these changes and their impact upon human performance, supplemental oxygen is required for aircrew members above 10,000-foot pressure altitudes for durations of more than 1 hour, for flights above 12,000-foot pressure altitudes for 30 minutes, or any time above 14,000 feet. Further information on oxygen requirements and training is contained in Army Regulation (AR) 95-1, *Flight Regulations*, and Training Circular (TC) 3-04.93, *Aeromedical Training for Flight Personnel*.

JUNGLE ENVIRONMENTS

3-23. Dense jungles limit the range and effects of weapons. Hot, humid air decreases aircraft lift, thus reducing the useful loads. Problems with aircraft corrosion increase, thereby increasing maintenance requirements. Safe landing areas are normally scarce.

ARCTIC ENVIRONMENTS

3-24. Cold temperatures increase lift capabilities and loads. However, snow and ice can adversely affect the aircraft's performance. Blowing snow limits visibility, especially during takeoff and landing. Extremely cold temperatures have an adverse effect on aircraft components. Because of the severity and frequency of changes in weather conditions in the arctic regions, climatic and weather conditions must be continuously monitored. Variations and changes in weather conditions that may impact flight operations must be reported quickly.

MARITIME ENVIRONMENTS

3-25. Air operations from naval vessels require extensive coordination with naval air operations to synchronize the location, sequence, timing of departing or returning flights, and long-range overwater and over-the horizon operations. Operating from naval vessels magnifies complications from adverse weather conditions. Increased airframe corrosion from saltwater exposure requires a freshwater source for aircraft washing. Increased frequency of aircraft washing may be necessary.

KEY OPERATIONAL CONSIDERATIONS

3-26. There are several key factors that must be addressed prior to the deployment of the SOAR to an operational area. The formalization of the C2 of the SOAR must be clearly articulated and understood. To assist in the C2 planning and communication, the SOAR will assign a liaison officer to the controlling command structure. The liaison officer will coordinate communications and mission requirements. The liaison officer will also act as the subject-matter expert on the SOAR's capabilities in an effort to exploit their full capabilities and ensure mission success.

JOINT OPERATIONS CENTER

3-27. The joint operations center is the functional activity that plans, coordinates, directs, and controls ARSOF operations in a designated area of operations. It performs the same functions as the conventional combat arms units' tactical operations center.

DECISIVE, SHAPING, AND SUSTAINMENT OPERATIONS

3-28. The SOAR can play an integral part in decisive operations and the shaping of the operational environment for exploitation by follow-on forces. The SOAR, with its advanced navigation systems, defensive electronics, and sustained high level of training, is especially suitable for deep-penetration operations well beyond the capabilities of standard Army aviation. Detailed information on decisive, shaping, and sustainment operations is contained in JP 3-0 and FM 5-0.

3-29. In sustainment operations, the SOAR can sustain itself with the assistance of the 528th SB(SO)(A) for limited periods of time. The ability for short-term sustainment is highly dependent upon the mission and area of operations. For prolonged operations and sustainment, the SOAR is dependent upon the Army Service component command and its subordinate tactical support command. Detailed information on logistics and sustainment is contained in JP 4-0, *Joint Logistics*; FM 4-0, *Sustainment*; and FM 3-05.140.

SOAR CHEMICAL, BIOLOGICAL, RADIOLOGICAL, AND NUCLEAR CAPABILITIES

3-30. The SOAR does not possess any organic capability to provide thorough decontamination. It can, however, conduct limited, hasty decontamination. The greatest chemical, biological, radiological, and nuclear threat is near the location of the highest concentration of troops. The mission-planning agent must be very mindful of known or suspected chemical, biological, radiological, and nuclear-contaminated areas and avoid them. The SOAR must coordinate with the Army Service component command chemical, biological, radiological, and nuclear decontamination element upon exposure of SOAR aircraft.

JOINT OPERATIONS AND RELATIONSHIPS

3-31. The SOAR in the CONUS is normally under the combatant command of the Commander, USSOCOM. When directed, the Commander, USSOCOM, provides CONUS-based SOAR forces to a geographic combatant commander. The geographic combatant commander normally exercises combatant command (command authority) of assigned and operational control of attached SOF through a TSOC commander. When a geographic combatant commander establishes and employs multiple joint task forces and independent task forces concurrently, the TSOC commander may establish and employ multiple JSOTFs to manage SOF and SOAR assets and accommodate joint task force or task force SO requirements. A JSOTF may also establish a JSOAC to place all aviation assets under a single command for the operation. A JSOAC allows for a single air commander to work with the SOTFs to maximize the support of available aviation assets to the SOTFs. SOAR typically operates as part of a JSOAC, but can also deploy in direct support to a SOTF depending on mission requirements. The geographic combatant commander, as the common superior, will normally establish support or tactical control command relationships between the JSOTF commanders and joint task force or task force commanders.

3-32. The Commander, USSOCOM, performs the role of lead combatant commander for planning, synchronizing, and (as directed) executing global operations against terrorist networks in coordination with other combatant commanders. When directed to execute global operations, the Commander, USSOCOM, can establish and employ JSOTFs as a supported commander. SOF that are used independently or integrated with conventional forces provide additional and unique capabilities to achieve objectives that may not otherwise be attainable. SOF are most effective when SO are fully integrated into the overall plan and the execution of SO is through proper SOF C2 elements employed intact, centralized, and fully responsive to the needs of the supported commander. SOF C2, coordination, and liaison elements normally provided to supported and supporting commanders are described in JP 3-05.

3-33. The standing joint force HQ core element is a staff organization that provides combatant commanders with a full-time, trained joint C2 element that is fully integrated into the combatant commander's planning and operations. The standing joint force HQ core element is staffed during peacetime to provide a core element of trained personnel. These personnel may serve as both a nucleus of key functional and C2 expertise and a foundation on which to build, through augmentation, the joint C2 capability for specific mission areas. The joint force HQ core element's principal roles are to enhance the command's peacetime planning efforts, improve operational area awareness for specific focus areas, accelerate the formation of a joint task force HQ, and facilitate crisis response by the joint force. The joint force HQ core element helps the combatant commander determine where to focus joint capabilities to prevent or resolve a crisis.

INTERAGENCY RELATIONSHIPS

3-34. A joint task force HQ is the operational focal point for participating in interagency coordination. During interagency operations, the joint task force HQ provides the structure for a unified effort. The level of a JSOTF HQ in the command structure of an operation dictates the JSOTF's roles and responsibilities in the interagency process. If the JSOTF is the senior or stand-alone joint task force, then the JSOTF assumes the primary responsibility as the focal point in the interagency process. Further information on interagency considerations is contained in JP 3-08, *Interorganizational Coordination During Joint Operations*.

3-35. Any control of ARSOF operating elements transferred to a foreign commander must include appropriate ARSOF C2 and liaison elements for direct C2 of the operating elements.

COORDINATION AND LIAISON

3-36. Liaison is that contact or intercommunication maintained between elements of military forces or other agencies to ensure mutual understanding and unity of purpose and action. Liaison facilitates communication of common-operational-picture-related information and execution information between different commands. In addition to passing information, liaison personnel can add meaning and context to information they send and receive. Liaison personnel can also expedite passage of relevant information that answers commander's critical information requirements and exceptional information. Detailed information on liaison activities and responsibilities is contained in JP 3-0 and JP 3-08.

Liaison Officer

3-37. A liaison officer is responsible for representing the commander at the HQ of another command to coordinate and promote cooperation between the two commands.

Liaison Fundamentals

3-38. Liaison helps reduce the fog of war through direct communications. It is the most commonly employed technique for establishing and maintaining close, continuous physical communication between commands. Commanders use liaison during operations and normal daily activities to help facilitate communication between organizations, preserve freedom of action, coordinate activities, and maintain flexibility. Liaisons answer operational questions and provide senior commanders with relevant information. Liaison ensures the commanders remain aware of the tactical situation by reducing risks and fratricide. Liaison activities augment the commander's ability to synchronize and focus combat power. They include establishing and maintaining physical contact and communication between elements of military forces and other government organizations, as directed. Liaison activities ensure—

- Cooperation and understanding between commanders and staffs of different HQ.
- Coordination on tactical matters to achieve unity of effort.
- Understanding of implied or inferred coordination measures to achieve synchronized results.

Chapter 4
Intelligence

INTELLIGENCE REQUIREMENTS

4-1. Because of the high risks associated with SOAR operations, their effectiveness depends largely on the ability of the S-2 to gather, produce, and disseminate detailed aviation-specific intelligence to mission planners in a timely manner. To successfully operate deep within hostile territory, the SOAR must avoid enemy detection. Avoiding enemy acquisition systems, therefore, is critical. Current intelligence on the location, status, and operating modes and frequencies of enemy acquisition and tracking systems is essential. The SOAR uses intelligence to plan routes and determine the needs and settings of aircraft survivability equipment. It also uses intelligence to determine the type of conventional and other government agency support it requires. SOAR mission planners use combat information and intelligence to plot infiltration and exfiltration routes and to recommend landing zones. Both general military and aviation-specific intelligence is vital to SOAR planners developing input to an SO mission planning folder. Typical information required to develop the SO mission planning folder is—

- Air order of battle.
- Air defense order of battle.
- Ground order of battle.
- Electronic order of battle.
- Naval order of battle.
- Imagery.
- Geospatial information.
- Climatology and terrain analysis.
- Recent threat activity and actions.

INTELLIGENCE ORGANIZATION

4-2. The only organic intelligence support in the SOAR and its subordinate battalions are their respective S-2 staffs. SOAR units have no other assigned intelligence assets. The organization and functions of the regiment and battalion are parallel, with the exception of the Intelligence Operations and Special Security sections found at the regimental level. The configurations of the regimental and battalion S-2 sections are depicted in Figures 4-1 and 4-2, page 4-2. The SOAR S-2 section is staffed by all-source, imagery intelligence, signals intelligence, and counterintelligence analysts; a special security officer; and an intelligence operations specialist. The SOAR S-2 staff can augment battalion S-2 sections, when required.

S-2 RESPONSIBILITIES

4-3. The regimental S-2 is a military intelligence major who is the primary staff officer responsible for all aspects of intelligence, counterintelligence, and security support in garrison. He coordinates weather support through collocated Air Force teams, manages the regiment map warehouse, provides macro intelligence for future operations, coordinates intelligence personnel augmentation for deployed units, and oversees counterintelligence support. The regimental S-2 pursues emerging technologies and coordinates required systems, organizations, and connectivity issues with USASOC. He manages the sensitive compartmented information facility through the special security officer and ensures efficiency and security of classified message traffic.

Chapter 4

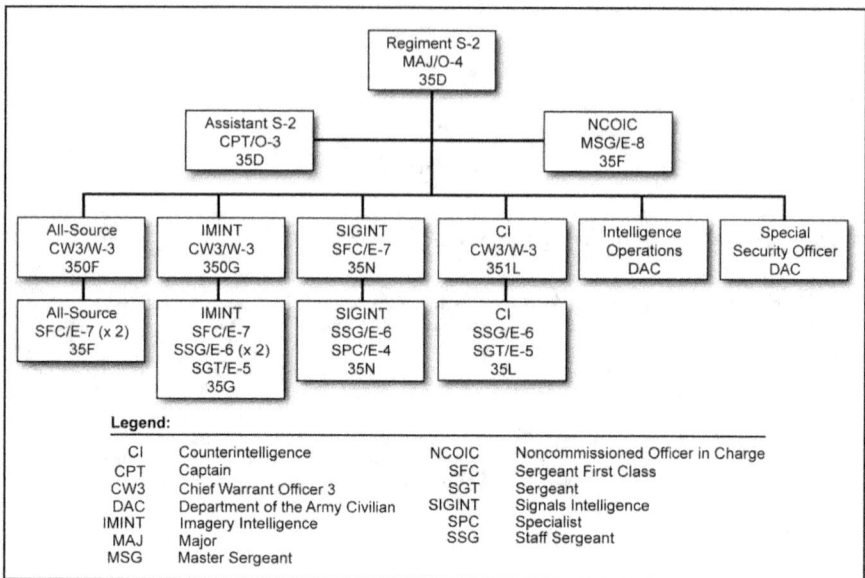

Figure 4-1. Intelligence organization of the SOAR

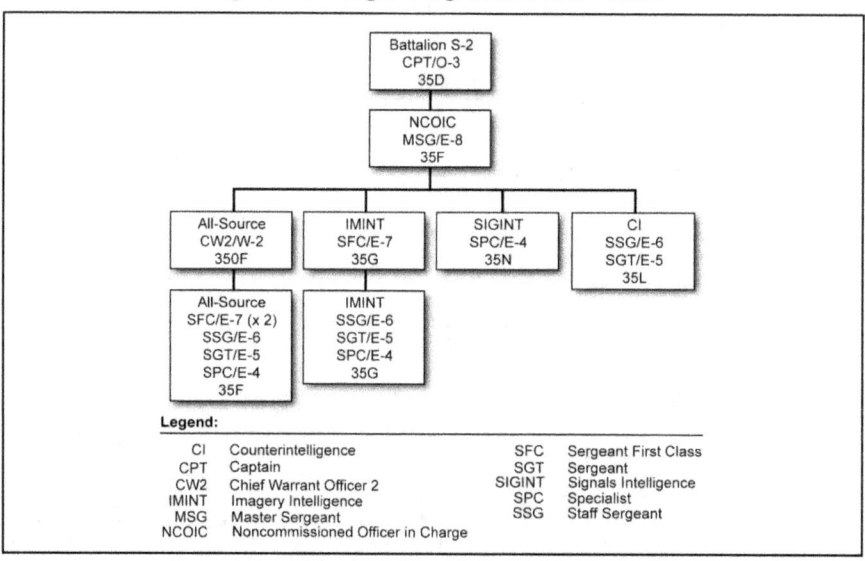

Figure 4-2. Intelligence organization of the SOAR battalion

Intelligence

ALL-SOURCE SECTION

4-4. The all-source analysts analyze intelligence, utilizing products from all intelligence disciplines. They study, analyze, integrate, and process threat information on targets identified in theater planning documents or for contingency operations in support of missions or training. The analysts track upgrades to country-specific ground and electronic order of battle, in addition to the modification and proliferation of surface-to-air systems. The analysts can augment battalion S-2 sections. They are led by an all-source intelligence technician and a senior intelligence analyst.

4-5. The all-source analysts' primary task is the identification and tracking of global threats for potential missions, using all available intelligence disciplines. Their focus is the modification and proliferation of surface-to-air missiles. They archive products, including overlays, situation updates, intelligence assessments, and weapons data research. The all-source technician coordinates with the SOAR's systems integration management office for the identification, research, and sharing of threat systems in relation to current aircraft survivability equipment capabilities. He also interfaces with the USASOC G-2 and G-8 with regard to new or updated analytical technology. He then coordinates the fielding of new analytical technology in the SOAR and its associated training for the all-source analysts.

SIGNALS INTELLIGENCE SECTION

4-6. The SOAR's signals analysts are responsible for providing signals intelligence and electronic intelligence analysis to aircrews and mission planners to ensure their aircraft flights are safe and undetected. The signals intelligence analysts are led by a senior signals intelligence analyst. The analysts can augment battalion S-2 sections when required.

4-7. The signals intelligence analysts provide near-real-time enemy communications intelligence and electronic intelligence situational awareness. They focus on premission analysis of both route threat and electronic detection capabilities of target countries. The SOAR signals intelligence section is able to produce the most current electronic order of battle and operations and activity clocks on associated electronic order of battle for SOAR areas of operations and interest. The analysts help determine current locations of enemy electronic and communications equipment, trends, and patterns of equipment and users. The senior signals intelligence analyst is responsible for researching, recommending, acquiring, and fielding new and updated technology for the SOAR through the systems integration management office and USASOC G-2 and G-8.

IMAGERY INTELLIGENCE SECTION

4-8. The SOAR's imagery intelligence section is the center of SOAR garrison imagery operations and supports forward-deployed units. The analysts are led by an imagery intelligence technician. There are also two National Geospatial-Intelligence Agency personnel—one imagery analyst and one geospatial analyst—who provide a conduit to the National Geospatial-Intelligence Agency for acquisition of special geospatial information and services products. When required, the regiment imagery analysts may be tasked to augment the battalions.

4-9. The SOAR imagery intelligence technician is responsible for pursuing emerging imagery technology and program upgrades for the regiment. When technology is fielded or updated, the imagery intelligence technician is responsible for researching, planning, and coordinating training for the SOAR imagery analysts. To support deployed units, the imagery analysts order, receive, manipulate, annotate, and disseminate imagery intelligence full-frame and chipped imagery products, both printed and electronic copies. These products include secondary imagery products, to include helicopter landing zone, drop zone, and noncombatant evacuation operation support; route assessments; terrain analysis; ingress and egress route analysis; line-of-sight studies; and point targets, such as embassies, ports, airfields, and training areas.

Chapter 4

INTELLIGENCE OPERATIONS SECTION

4-10. The SOAR intelligence operations section, consisting of a Department of the Army civilian, performs a full range of administrative, intelligence, and intelligence-related security functions. The section plans, develops, coordinates, and evaluates intelligence and security program goals and objectives, and assists with their implementation. The multidisciplined intelligence specialties include all-source operations, collection management, intelligence system requirements, and the security specialty of the Department of Defense Intelligence Information System. The intelligence operations specialist is responsible for performing the following functions:

- Develops all-source systems requirements:
  - Monitors the intelligence future and tactical exploitation of national capabilities programs to leverage developing techniques and technologies to increase intelligence productivity.
  - Identifies shortcomings in unit intelligence architecture and equipment capabilities.
  - Provides expertise, direction, and information regarding the acquisition of new intelligence systems.
- Oversees or otherwise conducts the duties of collection requirements, operations, and production management:
  - Generates, processes, and validates both improvised and standing all-source intelligence collection requirements.
  - Identifies and registers unit intelligence requirements to the Defense Intelligence Agency and manages the Defense Intelligence Agency accounts for the SOAR.
- Serves as the SOAR Department of Defense Intelligence Information System site information systems security manager with overall responsibility for the information systems security program:
  - Drafts appropriate facility standing operating procedures and chairs the site's Department of Defense Intelligence Information System configuration control board.
  - Has overall responsibility for coordinating and controlling access to worldwide intelligence databases.
- Establishes and maintains direct liaison, and coordinates unit intelligence and security requirements with USASOC; national and theater-level intelligence organizations; the military Services; combatant, regional, functional, and unified commands; and other appropriate agencies, as required.

COUNTERINTELLIGENCE SECTION

4-11. The SOAR counterintelligence personnel detect, evaluate, counteract, or prevent foreign intelligence collection, subversion, sabotage, and terrorism. They determine security vulnerabilities and recommend countermeasures. SOAR counterintelligence also supports deception operations, rear-area operations, predeployment site surveys, and counterintelligence support to personnel protection for all worldwide training exercises and real-world deployments. SOAR counterintelligence is led by a counterintelligence technician and includes a counterintelligence agent. Regimental counterintelligence personnel may augment the battalions.

4-12. The counterintelligence personnel support personnel protection efforts by providing multidiscipline counterintelligence products for SOAR operation plans. SOAR counterintelligence assets primarily focus on personnel protection for deployed SOAR elements. The counterintelligence personnel conduct counterintelligence liaison with U.S. and host-nation intelligence and law enforcement agencies, as required. In response to the situation and collection taskings from the S-2, the counterintelligence personnel plan, coordinate, and conduct counterintelligence liaison, overt collection, and displaced civilian debriefings. They recommend counterintelligence threat countermeasures, ensure compliance with intelligence oversight regulations, and continually assess the effectiveness of the base operations security countermeasures and base security plans. They also support the operations security plan by providing Subversion and Espionage Directed Against the Army briefings and limited investigations.

SPECIAL OPERATIONS WEATHER TEAM

4-13. The special operations weather team (SOWT) is a section within the S-3. However, the SOWT frequently performs a dual function, working with the S-2 section during the MDMP or joint operation planning process. The SOWT is responsible for accurate, detailed, and specific weather and environmental data. The SOAR is highly susceptible to the effects of weather, making it a critical aspect of mission planning. Severe local weather conditions can seriously degrade flight and target acquisition capabilities. Certain atmospheric conditions affect the propagation of aircraft noises, which impacts route selection. Even solar activity can degrade sensitive communication equipment. SOWTs provide timely weather support to the SOAR. Direct weather support includes, but is not limited to—

- Forecasts of general weather conditions and specific meteorological data elements as described in the 24-hour forecast.
- Geophysical information and climatic studies and analyses.
- Weather advisories, warnings, and specialized weather briefings, to include flight weather briefings of routes and objective areas.
- Lunar and solar illumination data, and light angle of incidence.
- Expected temperature variations along flight routes.
- Atmospheric factors, to include ceilings, visibility, air pressure, and wind conditions.
- Hazards to aircraft, such as icing, shear, and turbulence.
- Weather effects on timelines and schedules.
- Tidal data and water temperatures.

NONORGANIC INTELLIGENCE SUPPORT

4-14. Because the SOAR's organic intelligence support is limited, SOAR S-2s must be adept at leveraging support from the intelligence community. SOAR units normally operate as part of a larger SOF contingent, such as a JSOTF, JSOAC, or SOTF. Consequently, the SOAR S-2s should integrate their intelligence operations with those of other task force units. This integration is especially important when those units have a more robust intelligence support structure. Intelligence integration between the JSOAC, SOTF, SOAR, and reachback joint intelligence nodes is imperative to unmanned aircraft system operations. Tactics, techniques, and procedures, as well as task organization among intelligence elements, may be required to ensure relevant, timely, and accurate intelligence in support of the SOF mission.

COUNTERINTELLIGENCE AND HUMAN INTELLIGENCE SUPPORT

4-15. Although the SOAR has its own organic counterintelligence section, it may require augmentation from counterintelligence personnel assigned to higher echelons or liaison with counterintelligence personnel assigned to adjacent units. Support from nonorganic counterintelligence assets includes hazard reconnaissance, vulnerability assessments, and liaison support. The SOAR does not have any organic human intelligence assets. Therefore, the S-2 sections must request national or task force human intelligence assets for source operations, interrogation, liaison, document exploitation, and document and media exploitation. The organic SOAR counterintelligence personnel can perform tactical questioning and debriefings with nonorganic linguistic support and proper training.

IMAGERY INTELLIGENCE SUPPORT

4-16. The SOAR requires immediate access to current, detailed imagery products to plan all aspects of their precision assault missions. In addition to its organic imagery intelligence support, joint and national assets normally provide tactical imagery intelligence support to SOAR operations. Imagery intelligence sources include unmanned aircraft system feeds of radar, and photographic, infrared, and electro-optic imagery. The SOAR uses high-resolution imagery to develop detailed and sophisticated target folders. Imagery intelligence enables SOAR analysts to effectively plan their missions and contingencies, such as determining flight routes, clearance distances for helicopter landing zones, and locating hazards, such as wires or tall obstacles.

Chapter 4

SIGNALS INTELLIGENCE SUPPORT

4-17. The SOAR battalions have no organic signals intelligence support. When regiment support is not available, they rely on joint and national electronic intelligence databases and collectors to locate and identify electronic warfare emitters in the area of operations that could impact flight routes. Locating these emitters and their associated weapons systems is crucial to ensuring that the aircraft are able to fly undetected to and from their objectives.

INTELLIGENCE PREPARATION OF THE OPERATIONAL ENVIRONMENT

4-18. The SOAR uses the intelligence preparation of the operational environment process to support commanders and their staffs in the MDMP. The S-2 researches current intelligence databases and refines applicable reporting into aviation-specific intelligence products. These products are then posted to intelligence databases as a resource for aviation or aviation-supported units. The commander directs the intelligence effort by selecting and prioritizing intelligence requirements. These requirements support the commander in conducting and planning operations. Commander's critical information requirements are the information the commander needs to visualize the outcome of current operations. Commander's critical information requirements—

- Become priority intelligence requirements.
- Include information on both friendly and threat forces.
- Help develop intelligence analysis, which results in a written, graphic intelligence estimate that evaluates and portrays probable threat, friendly, and nonbelligerent third-party capabilities, as well as their respective vulnerabilities and probable courses of action.

4-19. Intelligence preparation of the operational environment is a cyclical process of intelligence analysis and evaluation that focuses on the assigned operational area and the forces expected to be operating in the area. It is the systematic, continuous process of integrating and analyzing data on the populace; threat, friendly, and nonbelligerent third parties; weather, climate, disease, and environmental threats; and terrain in a specific geographic area and operational environment.

4-20. Intelligence preparation of the operational environment is a graphic tool for presenting key characteristics of the operational environment. The essential difference between intelligence analysis for conventional forces and SOA is the level of detail required in each step of the analysis process for SOA missions. The SOAR normally follows a five-step process explained in the following paragraphs.

STEP 1 – EVALUATE THE OPERATIONAL ENVIRONMENT

4-21. The process begins with an area evaluation. The assessment of the area of operations includes the overall nature of the friendly and enemy forces, and the operational environment. In this step, the S-2 determines and answers requirements for weather, climate, and terrain.

4-22. The S-2, working closely with the unit surgeon, develops detailed threat intelligence on disease and the environment of the target area and intermediate staging areas. The surgeon uses intelligence to develop appropriate countermeasures to the medical threat; for example, immunizations, chemoprophylaxis regimes, and preventive medicine countermeasures. These measures help decrease the loss of aircrew availability due to disease or nonbattle injury.

4-23. During the area evaluation for SOA, the S-2 evaluates threat ground, air, and naval forces expected to operate within the operational environment. This evaluation includes routes to and from the target. These evaluations are useful in determining the capabilities of the forces in relation to the weather, terrain, and friendly mission. Particular attention is on air bases, including ships carrying aircraft, refueling points, landing zones, drop zones, minefields, and air defense weapons, radar, and other sensors operating within the operational environment.

Intelligence

STEP 2 – ANALYZE THE TERRAIN

4-24. The purpose of terrain analysis is to reduce the uncertainties and effects of natural terrain and man-made obstructions, and to assess the effects of the population on military operations. The SOAR conducts in-depth studies of each course of action flight route during the terrain analysis step. This study is imagery-intensive because target areas are often inaccessible by ground or are behind enemy lines. The SOAR operates primarily at night and under limited visibility. Identifying and measuring terrain features is, therefore, often critical to the safety of the aircrew and assault forces, and to the success of the overall mission.

4-25. Highly detailed imagery is necessary for precise preflight planning. Because the SOAR flies long infiltration and exfiltration routes, critical terrain consists of identifiable reference points that aid in navigation, as well as terrain features that can mask friendly aircraft from detection.

4-26. The SOAR's terrain overlays depict all the obstacles to flight, reference points, checkpoints, masked areas, and danger zones. These factors are useful in determining the best flight routes to the target. The S-2 and engineers use the modified combined obstacle overlay to determine flight routes for infiltration and exfiltration. This overlay is the basis for input to the SOAR's requirements section of the target intelligence package. When preparing to support direct action and special reconnaissance ground missions, SOA terrain analysis must surpass the detail normally developed by the ground force terrain analysis. Because of the low altitudes SOAR aircraft fly, features that are obstacles to ground forces may also be obstacles to SOA. SOA terrain analysis, for example, requires the S-2 and engineers to provide heights of buildings, poles, trees, wires, streets, open fields, and anything else that could be in a landing zone or pickup zone, or be an obstacle to flight during insertions and extractions. The theater terrain team may provide terrain support if time permits.

STEP 3 – ANALYZE THE WEATHER

4-27. Detailed weather analysis is a necessity. Generic weather summaries for a country are insufficient for SOAR elements. Development of the best flight routes requires weather patterns for each geographical region and a comparison of the terrain data. Weather data that may have a negligible impact on conventional Army aviation assets may be critical for night infiltration operations in denied areas.

4-28. Within the target area, last-minute weather conditions or possible environmental phenomena may be critical elements to the target analysis and overall mission planning. The following weather aspects and phenomena are a partial list of conditions that have a direct impact upon the SOAR's mission planning and execution:

- Moon illumination and angle are important for flight operations with night vision goggles.
- Visibility, wind speed, and wind direction can significantly affect light-helicopter operations.
- Conditions of sand or snowstorms in a moderate wind, loose rock and gravel in a high wind, and sudden brownout or whiteout, as well as icing, can render SOA operations ineffective.
- Sea and water conditions, to include temperatures, are also important to know for survivability and for combat search and rescue operations when flying over water.

STEP 4 – EVALUATE THE THREAT

4-29. Threat evaluation is a detailed study of threat forces and their composition, organization, tactical doctrine, weapons, equipment, reaction times, and supporting systems. Threat evaluation determines threat capabilities and limitations and the way the threat fights if unconstrained by weather and terrain.

4-30. Threat evaluation for SOA is often difficult and complex because of the environment in which SOA aircraft work. Many of the systems presenting a threat to SOA do not affect conventional commanders. Because the mission requirements for SOA in direct action and special reconnaissance missions are to infiltrate and exfiltrate undetected, the primary threat is anything that can detect and report aircraft movements.

Chapter 4

4-31. During threat evaluation, the SOAR S-2 also examines enemy communications links. A threat that can detect the mission aircraft but cannot report its presence in a timely manner is not a major concern to the SOAR element. On the other hand, a lone rifleman with a radio or a telephone can compromise a SOAR mission. Terrain masking is a critical factor in determining the threat's detection and reporting capabilities.

4-32. Threat evaluation of the SOA target site itself is also complex and goes well beyond the usual order of battle available for the target area. SOA elements must know almost as much as the ground element about security forces. They also need information on aerial patrols, reaction forces, lighting at the target, and the enemy equipment and capabilities believed to be in the target area or along flight routes. When the SOAR and other ARSOF work together, close coordination or even consolidation of their threat evaluations is necessary to reduce planning times and risks.

STEP 5 – INTEGRATE THE THREAT

4-33. The integration of all the factor analyses in the preceding steps occurs during this final phase of the process. It presents a total picture to the S-2, the commander, his staff, and the assault force. Templates play a key role in presenting this picture to the commander.

4-34. The S-2 uses situation templates to show how threat, friendly, and nonbelligerent third-party forces might operate and communicate within the constraints of specific meteorological conditions and sociopolitical geography. The situation template is basically a doctrinal template that identifies the critical threat, friendly, and nonbelligerent activities and their locations. It also provides a basis for situation and target development and high-value target analysis.

4-35. Event templates show locations where critical events and activities are to occur and where critical targets and opportunities are to appear. The S-2 uses the event template to predict time-related events within critical areas. It provides a basis for collection operations predicting threat, friendly, and nonbelligerent third-party intentions, and locating and tracking high-value targets.

4-36. The S-2 and the S-3 use a decision support template to show decision points keyed to significant events and activities. The template is the intelligence estimate in graphic form. It identifies critical events and human activities related to time and location that may require a tactical or operational decision by the commander.

4-37. During this step, the S-2 must also integrate pertinent threat data into situational and event templates that affect the choice of flight routes and modes of infiltration and exfiltration. This phase is the final step in the development of the target folder. The threat as it relates to flight route options is the final factor in deciding which operation provides the best route. Threat integration helps determine the best approach and final assault paths into the target area. It provides the necessary visual products for the integration of coordinated assault fires, if needed. Situation templates are important for SOA only if the element must perform sustainment operations.

4-38. This step integrates threat data with the target terrain data, including the location and dimension of every structure and obstacle. The final product—

- Depicts the best insertion or extraction points.
- Identifies targets for destruction by support attack helicopter fire.
- Helps reconcile multiple flight routes in limited airspace.

Finally, threat integration results in the decision support template for SOA, which depicts—

- Overlays of the air defense zone.
- Reaction times of threat aircraft.
- Combat radio overlays.
- Terrain-masking overlays.
- Other threat courses of action.

4-39. The decision support template shows the best flight route and course of action on the target under varying conditions. It depicts terrain, obstacles to flight, routes, landing zones, pickup zones, alternate

Intelligence

pickup zones, orders of battle, targets or target areas of interest, and operational timelines. In a sustained direct action or special reconnaissance mission, personnel use named areas of interest along with target areas of interest when SOA aircraft perform interdiction missions.

UNMANNED AIRCRAFT SYSTEMS

4-40. Historically, unmanned aircraft systems have been used as platforms for the collection of intelligence, surveillance, and reconnaissance. There has been a dramatic increase in the use of unmanned aircraft in the tactical role. Some have been outfitted with offensive weaponry; some have provided laser designation; and hand-launched systems have located snipers, improvised explosive devices, mortar firing points, and fleeing insurgents. Unmanned aircraft systems have enhanced the situational awareness of unit-level commanders by providing more accurate and immediate battle damage assessment. The adaptability, versatility, and cost effectiveness of unmanned aircraft systems continue to expand the commander's warfighting capability and have become indispensable to successful joint combat operations.

UNMANNED AIRCRAFT SYSTEM COMPONENTS

4-41. Unmanned aircraft systems consist of several components that are common among all system groups. The components are the unmanned aircraft, payloads, communications, control equipment, support equipment, and the human element. Tasking, processing, exploitation, and dissemination, while not technically part of the unmanned aircraft system, is critical to optimize the capabilities of unmanned aircraft systems with mission payloads.

EMPLOYMENT

4-42. Within SO, unmanned aircraft systems are predominantly employed as intelligence, surveillance, or reconnaissance assets, although the MQ-1C fielded to the SOAR is a multipurpose unmanned aircraft system that is weapons-capable. Planning and employment of these systems in roles such as intelligence, surveillance, and reconnaissance; air interdiction; close combat attack; close air support; and personnel recovery are the same as manned fixed-wing and rotary-wing with additional considerations for C2 sources and means, target acquisition, and strike.

SPECIAL OPERATIONS UNMANNED AIRCRAFT SYSTEMS

4-43. Unmanned aircraft systems are organic to the SOAR and the SF groups. Current systems within ARSOF include the MQ-1C (Grey Eagle), the RQ-7B (Shadow), the RQ-11B (Raven), and the Wasp Block III (Table 4-1).

Table 4-1. Army special operations unmanned aircraft system platforms and payloads

Name	Mission Design Series	Electro-Optical Infrared	Full-Motion Video	Remote Video Terminal	Laser Range Detector/Laser Range Finder	Infrared Pointer	Synthetic Aperture Radar	Ground Moving Target Indicator	Signals Intelligence	Communication Relay	Weapons/Cargo
Grey Eagle	MQ-1C	X	X	X	X	X	X	X	X	X	Hellfire
Shadow	RQ-7B	X	X	X	X	X					
Raven	RQ-11B	X	X	X		X				X	
Wasp Block III		X	X	X							

Chapter 4

160th Special Operations Aviation Regiment

4-44. While not yet completely fielded, the SOAR is employing the MQ-1C (Grey Eagle) (Extended Range/Multipurpose [ERMP]). This unmanned aircraft system is capable of intelligence, surveillance, and reconnaissance; infrared pointer; and communications relay. It is also weapons-capable with the Hellfire missile. The MQ-1C is a Group IV system, requiring significantly more maintenance and support infrastructure than the systems at the SF groups or Ranger Regiment. The MQ-1C can provide up to 36 hours of operation within a 648-nautical-mile radius and is capable of supporting all of the ARSOF core activities.

Special Forces Groups

4-45. There is one RQ-7B (Shadow) platoon at the group level (Figure 4-3). Each Shadow can provide 18 continuous hours of surveillance in a 24-hour period within a 68-millimeter radius. From the SF group, they are tasked to support SF operations as needed. In addition to the Shadow platoon, an RQ-11B (Raven) is distributed with two Raven systems at group level, and two systems at each battalion and company. The Raven is a hand-launched system that can provide 60 to 90 minutes of surveillance at 8 to 12 kilometers distance. The Wasp Block III is a hand-launched micro-unmanned aircraft system distributed at each Special Forces operational detachment A (SFODA) and can provide up to 45 minutes of surveillance at up to 5 kilometers distance. Figures 4-4 through 4-7, pages 4-11 and 4-12, depict the MQ-1C (Grey Eagle), the RQ-7B (Shadow), the RQ-11B (Raven), and the Wasp Block III.

75th Ranger Regiment

4-46. As with the SF group, there is one RQ-7B (Shadow) platoon at regimental level. The Shadow platoon is tasked as needed to subordinate units.

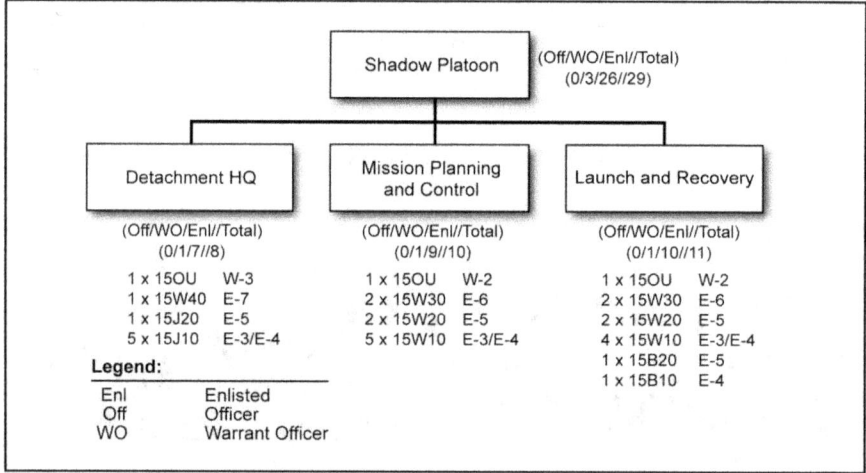

Figure 4-3. Shadow platoon organization

Intelligence

Figure 4-4. MQ-1C (Grey Eagle)

Figure 4-5. RQ-7B (Shadow)

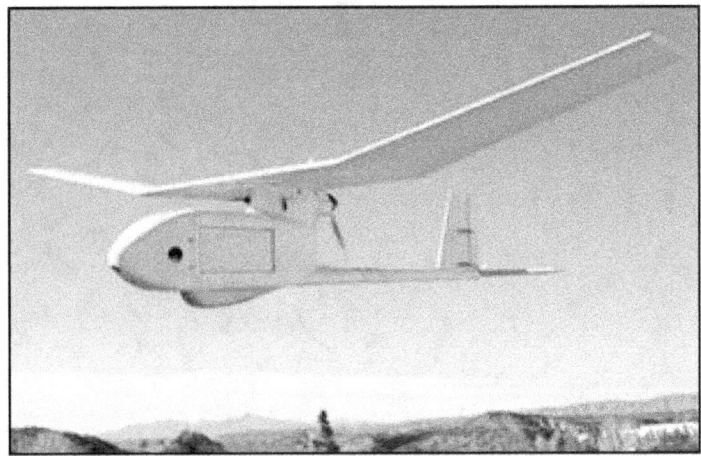

Figure 4-6. RQ-11B (Raven)

Chapter 4

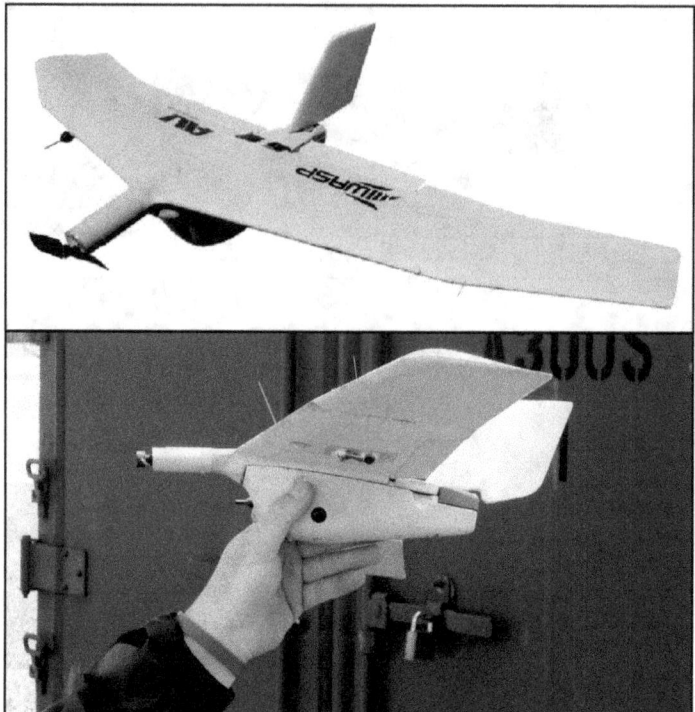

Figure 4-7. Wasp Block III (overhead and side views)

Chapter 5
Communications

COMMUNICATIONS ELEMENTS

5-1. The SOAR S-6 is located in the regiment headquarters and headquarters company. In addition, each aviation battalion has an S-6 and a communications-electronics section. Detailed information on ARSOF communications support is available in FM 3-05.160, *Army Special Operations Forces Communications System*.

REGIMENTAL S-6 AND COMMUNICATIONS-ELECTRONICS SECTION

5-2. The regimental S-6 is the principal staff officer for all matters concerning signal operations, knowledge management, network operations, and information security. The S-6 heads a small communications-electronics section that plans signal operations, prepares the signal annex to the operation orders, recommends employment of the SOAR's communications assets, and supports electronic warfare operations focusing on electronic protection. In addition, the regimental S-6 includes the regiment's frequency manager and the communications security custodian. The section mans a fixed-site base station at the regimental HQ. The regimental S-6 has a separate automated data processing section that conducts network operations at all classification levels. This section is led by the regimental automation officer under the regimental signal officer.

BATTALION SIGNAL ELEMENTS

5-3. The battalion S-6 is the principal staff officer for all matters concerning signal operations, automation management, network management, and information security at the battalion level. His duties are similar to the regimental S-6 with the exception of communications security. Battalion S-6 personnel can deploy with a forward element or augment the fixed-site base station at the regimental HQ. All the separate forward-deployed companies have a small communications section of several Soldiers.

COMMUNICATIONS CAPABILITIES AND EQUIPMENT

5-4. The SOAR's communications must support air-to-air and air-to-ground covert aircraft communications for C2, mission deconfliction, and mission support to SOF units. Figure 5-1, page 5-2, depicts the communications networks of a deployed SOAR task force. The location of the JSOACC depends on the command relationship between the supporting and supported units.

5-5. Capabilities of the regimental HQ's fixed-site base station include ultrahigh frequency single-channel satellite communications, super-high frequency multichannel tactical satellite, and high frequency, very high frequency, or frequency modulation (FM) radio. The regiment can also establish a forward element with the same capabilities if augmented by the ARSOF's 112th Signal Battalion (A) or joint signal elements.

5-6. The SOAR's aircraft employ multiband satellite communications, single-channel ultrahigh frequency satellite communications, high frequency burst and data, and amplitude modulation (AM) and FM line-of-sight radios that are interoperable with supported SOF units. The SOAR has the organic capability to terminate voice, data, and video services for short-duration missions. It does not have organic "broadband" assets for sustained operations, and requires augmentation by the 112th Signal Battalion (A), joint communication elements, or a theater-based signal unit.

5-7. The regiment employs the special operations forces deployable node-medium (SDN-M) satellite terminal establishing the necessary communications for a SOAR HQ at remote locations. This small

Chapter 5

communications package supports initial entry or a battalion-minus-level HQ. This situation also requires augmentation from the 112th Signal Battalion (A) or a theater-based signal unit. The minimum communications support package for establishing a remote site includes data (Non-Secure Internet Protocol Router Network, Secret Internet Protocol Router Network, and generally, Joint Worldwide Intelligence Communications System connectivity, depending on mission), voice (Defense Switched Network), and video teleconferencing (Secret Internet Protocol Router Network and Joint Worldwide Intelligence Communications System). The aggregate minimum data requirement for any SOAR C2 element is 2 megabytes per second because of the imagery requirements associated with SOAR operations.

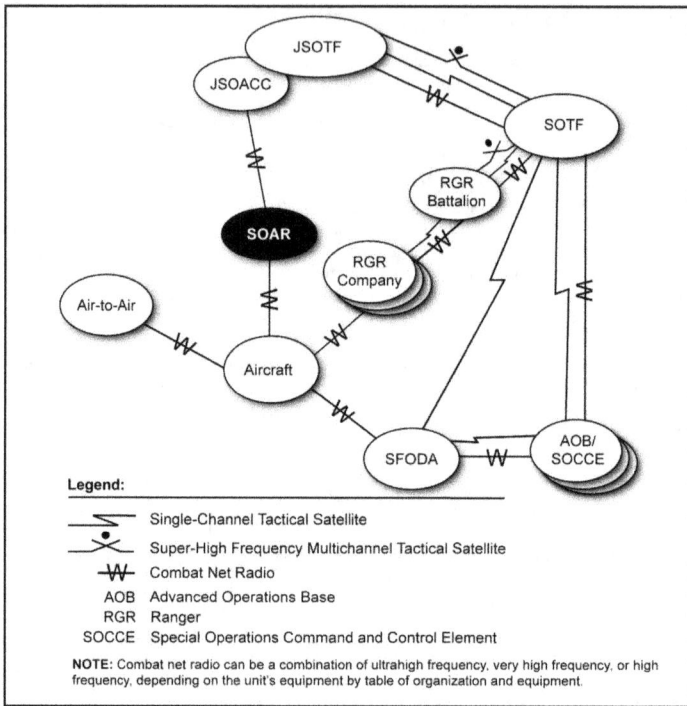

Figure 5-1. SOAR communications

CONCEPT OF EMPLOYMENT

5-8. The SOAR units plan, conduct, and support SO missions unilaterally or jointly in all theaters of operation, and at all levels of conflict. To accomplish this mission, SOAR units are task-organized according to the supported unit, the theater of operations, and expected missions. These organizations normally form around a battalion. The SOAR task force supports a specific mission based on the mission variables. Its size can range from as small as two or three helicopters to support a team, to a battalion-plus-sized element supporting a battalion.

5-9. The SOAR may be task-organized under a joint force commander, joint force air component commander, JFSOCC, JSOACC, JSOTF, or SOTF. Doctrinally, the SOAR is never under operational control or tactical control of conventional forces. It is normally under operational control of a JSOACC or

Communications

SOTF. With proper personnel and equipment augmentation, the SOAR battalion commander and his staff could serve as a JSOACC. When two or more battalions are required in the theater, the regimental commander could serve as the JSOACC. The following clarifies the SOAR's C2 relationships:

- *Joint force commander.* The SOAR normally does not come under the direct control of a joint force commander, as such control may encumber the SOAR's ability to support SOF.
- *Joint force air component commander.* The joint force air component commander is the Service component commander who has the preponderance of air assets and the ability to C2 these assets. The joint force commander designates the joint force air component commander. When under operational control of the joint force air component commander, the SOAR loses its identity as a SOF asset and does not support SOF ground forces.
- *JFSOCC.* The JFSOCC is the functional SOF component commander with the preponderance of SOF and the requisite C2 capabilities within the joint force. Operational control of the SOAR by the JFSOCC is usually delegated to the JSOACC or placed under operational control of the SOF ground commander.
- *JSOACC.* The JSOACC is the JFSOCC Service commander who either has the preponderance of the aviation force or is most capable of controlling special air operations within a given environment. The JFSOCC designates the JSOACC. He is the single air manager. The JSOACC allocates the SOAR to SOF missions as required to support JFSOCC missions.
- *SOTF.* If the SOAR is under the control of a SOTF, the command relationship is normally operational control.

5-10. Liaison support is on an as-required basis. The SOAR's liaison officers can perform tasks in the following organizations:

- *SOLE.* The SOLE is a special staff provided to the joint force air component commander or appropriate Service component air C2 element. The purpose of the SOLE is to synchronize SOF air and ground efforts with joint air operations. The SOLE chief works directly for the JFSOCC. The SOLE effects any required coordination with the SOAR unit and conversely provides the conduit for the SOAR unit to coordinate requirements, airspace, and deconfliction.
- *Joint search and rescue center.* The SOAR's support to the Joint Search and Rescue Center occurs when the SOLE and Joint Search and Rescue Center are not collocated at the joint force air component commander. Normally, the SOAR's liaison officer at the SOLE can cover both requirements when collocated. The liaison officer will coordinate and deconflict SOAR assets allocated to the Joint Search and Rescue Center force.
- *JFSOCC.* The mission, the number of aircraft, and the intensity of the operation determines the configuration of the SOAR's liaison officer cell. The SOAR's liaison officer coordinates mission support with the supported unit and deconflicts airspace requirements. The SOAR's liaison officer also provides expertise to the SOTF commander on the tactical employment of aviation assets.
- *SOTF.* Liaison officer support to the SOTF is according to the mission and C2 relationship between the SOTF and the SOAR unit. The liaison officer is primarily responsible for providing tactical and technical advice, and facilitating coordination with the supporting SOAR's assets. The higher HQ that retains C2 of the SOAR provides airspace coordination and deconfliction. If the SOAR is under operational control of or attached to the SOTF, then a liaison officer cell forms to provide advice and to effect the airspace coordination and deconfliction. Establishing the links between the airspace control authority and the liaison officer cell is very important in properly integrating the SOTF's airspace requirements within the area of operations.
- *Special operations command and control element (SOCCE).* The SOCCE should address its aviation concerns to the SOAR's liaison officer in the SOTF. The mission of the SOCCE is to synchronize SOF missions with the supported conventional force command post. A SOAR liaison officer should be available at limited times at the SOCCE level. Generally, the SOCCE requests a liaison officer when—
  - The SOCCE has operational control of the majority of the forward-deployed SOF.
  - SOAR liaison officer support at the SOTF or SOLE is not in a position to enhance coordination in the SOCCE's area of operations.

Chapter 5

5-11. The SOAR's communications must support covert air-to-air and air-to-ground aircraft communications for C2, mission deconfliction, and mission support to SOF units. Figure 5-2 depicts the C2 networks necessary to support the SOAR structure. The location of the JSOACC depends on the command relationship between the supporting and supported units.

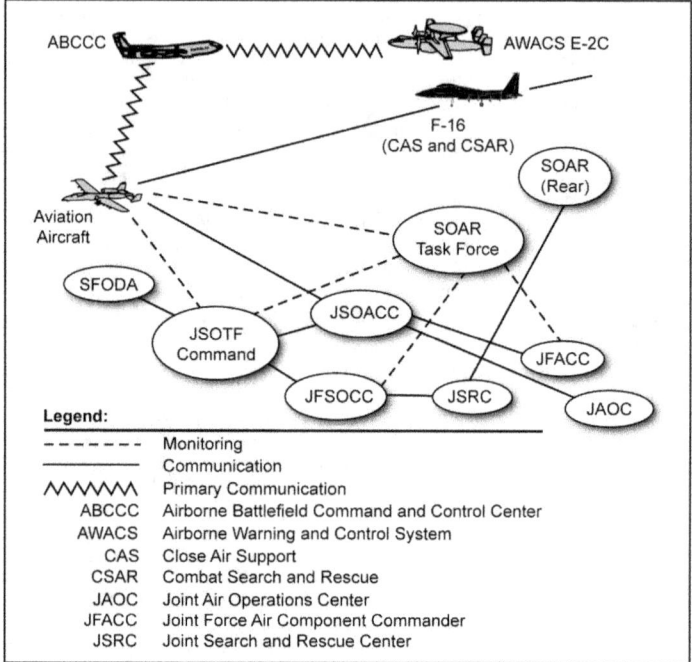

Figure 5-2. SOAR communications network architecture

5-12. The SOAR's aircraft employ multiband satellite communications, single-channel ultrahigh frequency satellite communications, high frequency burst and data, and AM or FM line-of-sight radios. These radios are interoperable with supported SOF units. Figure 5-3, page 5-5, illustrates the SOAR's communications systems connectivity.

5-13. The SOAR requires Joint Worldwide Intelligence Communications System, Secret Internet Protocol Router Network, Non-Secure Internet Protocol Router Network, video teleconferencing, and voice services. Circuits will often need to be extended from the main tactical operations center to aircraft maintenance and staging areas, often over an airfield. Planners should always plan for circuit extensions when planning SOAR support. The SOAR deploys specialized flight simulators for mission rehearsals that have unique computer network requirements and use significant bandwidth when downloading imagery and simulation data from their home station.

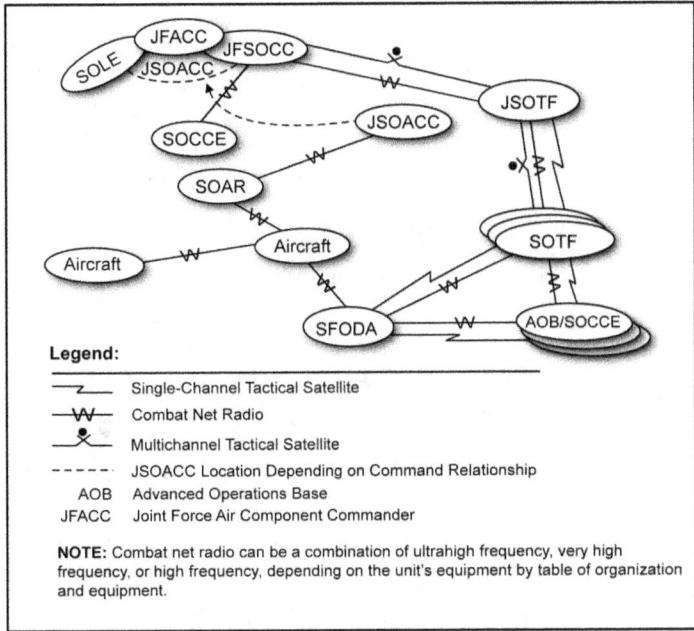

Figure 5-3. SOAR communications systems connectivity

UNMANNED AIRCRAFT SYSTEM CONNECTIVITY

5-14. Unmanned aircraft may be controlled by a centralized or joint C2 node, such as the joint air operations center. They can also be controlled by a forward tactical C2 node, such as a ground maneuver tactical operations center, an airborne warning and control system, or a joint terminal attack controller. The unmanned aircraft system may also operate independently of other C2 nodes by communicating directly with other fixed- and rotary-wing assets. There are three basic C2 options:

- The joint air operations center controls the unmanned aircraft and receives the video feed. The joint air operations center passes information gathered from the video to the tactical C2 node.
- The tactical C2 node receives the unmanned aircraft video and controls the mission without going through the joint air operations center. Terminal control in close air support can be managed in this C2 option. Ground maneuver forces can use the unmanned aircraft to conduct targeting for indirect fires (for example, mortars or artillery).
- The unmanned aircraft operator independently finds targets of opportunity and communicates directly with other airborne assets or appropriate ground elements (battalion tactical operations center or command vehicle) to provide targeting data. This option is normally used when conducting air interdiction and the unmanned aircraft and additional assets have the authority to strike targets in a designated area in accordance with interdiction rules of engagement.

5-15. Within ARSOF, missions for the MQ-1C (Grey Eagle) and RQ-7B (Shadow) are tasked through the combined JSOTF military intelligence section within the operations center. The ground control station is usually at or near the combined JSOTF, where a local area network is the preferred means of communication between the operations center and the ground control station. Multichannel tactical satellite can also be used if a local area network is not available.

Chapter 5

5-16. While flying the mission, the Grey Eagle or Shadow transmits data to the combined JSOTF, for further tasking, processing, exploitation, and dissemination. A SOTF or advanced operations base can also receive data directly from the unmanned aircraft. SFODAs do not have the capability to receive data directly from the Grey Eagle or Shadow unless they are within 25 kilometers line of sight and equipped with a remotely operated video enhanced receiver or the One System Remote Video Terminal, with the correct receiver configuration loaded. The data must be forwarded to them from the combined JSOTF, SOTF, or advanced operations base. Figures 5-4 through 5-6, pages 5-6 through 5-8, show typical unmanned aircraft system connectivity.

5-17. The RQ-11B (Raven) is a small tactical unmanned aircraft system and is most often used by the SFODA or advanced operations base. The ground control station is typically collocated with the unit receiving the data, and the ground control station terminal receives the data for immediate use by the tactical element.

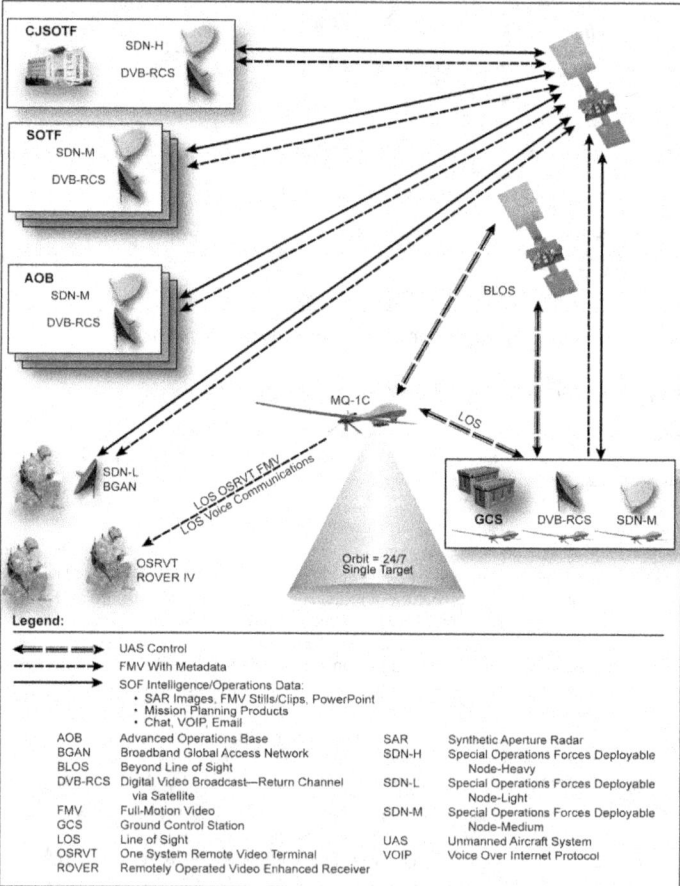

Figure 5-4. MQ-1C (Grey Eagle) connectivity

5-6 FM 3-76 28 October 2011

Communications

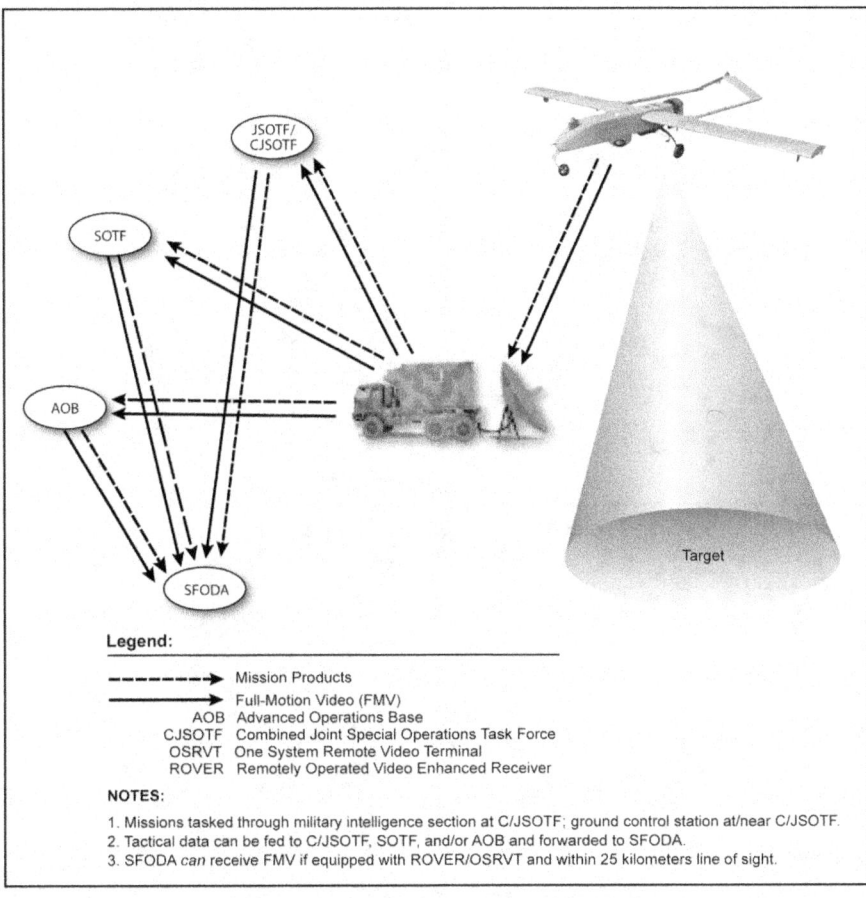

Figure 5-5. RQ-7B (Shadow) connectivity

Chapter 5

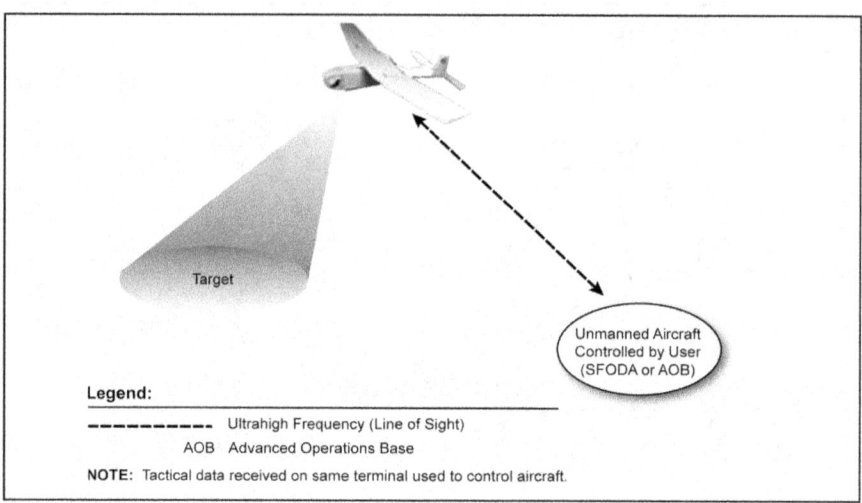

Figure 5-6. RQ-11B (Raven) connectivity

Chapter 6
Fires

TERMINOLOGY

6-1. In the execution of SOAR missions, the need to plan and integrate fire support, fires, suppression of enemy air defenses, and joint suppression of enemy air defenses into the overall operational plan is critical. The failure to effectively plan and coordinate these fire support supporting functions can lead to mission failure and cause unacceptable loss of equipment and highly trained personnel that cannot be readily replaced. Because of its advanced avionics, aircraft, and aircrew capabilities, the SOAR can effectively support fire support and fires missions in an effort to shape the operational environment.

6-2. The terms "fire support" and "fires" encompass several terms and concepts that include, but are not limited to, lethal and nonlethal effects, suppression of enemy air defenses, and joint suppression of enemy air defenses. These terms as articulated below are defined in JP 1-02, *Department of Defense Dictionary of Military and Associated Terms*:

- *Fire support* is defined as fires that directly support land, maritime, amphibious, and special operations forces to engage enemy forces, combat formations, and facilities in pursuit of tactical and operational objectives.
- *Fires* is defined as the use of weapon systems to create a specific lethal or nonlethal effect on a target.
- *Suppression of enemy air defenses* is defined as that activity which neutralizes, destroys, or temporarily degrades surface-based enemy air defenses by destructive and/or disruptive means.
- *Joint suppression of enemy air defenses* is defined as a broad term and activity that includes all suppression of enemy air defense activities provided by one component of the joint force in support of another.

COORDINATION AND PLANNING

6-3. Suppression of enemy air defenses in support of the SOAR may be necessary to penetrate and exit enemy territory during the conduct of SOF missions. Enemy reaction usually occurs because of increased activity in their operational areas. Therefore, careful and thorough planning for penetration and exiting from enemy territory is necessary to prevent disclosure of the targeted areas. Ideally, joint air defense suppression support should be congruent with other operations or activities.

6-4. Air defense suppression activities can also distract or act as diversionary action against enemy defenses from the actual planned routes or sequence of SOAR operations. This action is referred to as a feint. A feint is a limited objective attack. It is a show of force intended to deceive the enemy and draw attention away from the main attack. A feint must be conducted with sufficient strength and composition to cause the desired enemy reaction. Feints must appear real; therefore, some contact with the enemy will most likely be required.

6-5. Air defense suppression is a temporary activity. It does not have a long-term effect unless it is part of a major extended air defense suppression operation. Some SOAR mission requirements may be dedicated to direct suppression support. When this direct support is necessary, the timing of the suppression must take advantage of the previous operational suppression planning, while not jeopardizing the SOAR mission. The request must specifically state the greatest need for suppression, including where, when, and type of support required. Normally, full coverage of an operation is impossible because of limited, dedicated suppression assets. Support may also be split, such as between tactical air and naval and artillery fires, and be for different areas or targets or at different times.

6-6. Like tactical air, requests for air defense suppression go through the theater joint force air component commander. Suppression planning occurs at the joint air operations center level. Although local air defense suppression often supports high-priority missions, planning is normally on a theater or area basis rather than a single mission requirement. Tasking assets to fill a SOAR suppression request is, therefore, normally a part of a comprehensive suppression support package and plan. To compete for scarce assets, suppression requirements must be on time within the theater's air tasking order cycle. An exception to the air tasking order cycle requirement may be achieved from the development of a predesignated kill box in accordance with FM 3-09.34, *Multi-Service Tactics, Techniques, and Procedures for Kill Box Employment*.

6-7. The JSOAC refers to the commander, staff, and assets of an SO air component of a subordinate unified command, a JFSOC, or a JSOTF. A JFSOCC or JSOTF commander may establish a JSOAC. It is established as a functional component within a joint SO force to control SOA assets. The JSOACC is the commander within a joint SO command responsible for planning and executing joint SO air activities, to include the responsibility to coordinate, allocate, task, control, and support the assigned joint SOA assets. The establishing SOF commander (JFSOCC or JSOTF commander, as appropriate) normally exercises operational control of joint SOA through a JSOACC. However, there are also circumstances where the SOF commander may elect to place selected SOA assets under separate control. A JSOAC may be subordinate to a single JSOTF or separate—tasked to support the activities of multiple JSOTFs within a JFSOC. A JSOAC may be a standing organization or can be formed in response to a contingency or other operation.

ASSETS AND TECHNIQUES

6-8. Air defense suppression may be achieved by fires, electronic means, and by airborne or surface-based systems. For example, artillery or tactical air can silence most defenses, and airborne jammers can degrade hostile acquisition systems. Combining these assets increases the total effect of suppression capabilities. Normally, air defense suppression is part of an overall operation that includes other tactical air, naval and artillery fires, and ground suppression activities. Support may not be obvious unless the supported unit knows that air defense suppression is taking place. Limited numbers of lethal and nonlethal suppression systems may restrict the availability of suppression, particularly under quick-reaction requirements. In general, operations should include the planning for employment of suppression and kill box procedures on all known targets, and infiltration and exfiltration routes.

COMBAT AIR PATROLS

6-9. The SOAR may request dedicated air defense requirements or combat air patrols over enemy territory before mission execution. Like air defense suppression, combat air patrols attract the attention of the enemy. To prevent compromising SOAR activities, units must use combat air patrols judiciously. Combat air patrols may serve as a barrier between enemy air bases and the SOAR's overall mission and actions on the objective. A moving combat air patrol can provide cover near the SOAR's infiltration and exfiltration routes, or perform combat air patrol cover sweeps ahead of SOAR aircraft. In any case, close coordination with the supporting unit is an absolute necessity.

ATTACK HELICOPTERS

6-10. SOAR attack helicopters typically serve as armed escort and support helicopter assault forces during infiltration, exfiltration, and other contingency operations. They also act as an extension of the ground force commander's fire maneuver element, providing precision on-call fire support. SOAR attack helicopters can perform a variety of missions, to include preassault fires, on-call precision fire support, reconnaissance, and autonomous direct action against a known threat or target. Appendix B includes information on the SOAR attack helicopter call-for-fire format.

6-11. The SOAR can perform limited suppression missions with its organic attack helicopters. The decision to use SOAR assets in a suppression mission must undergo in-depth execution planning and risk-versus-reward analysis using the MDMP.

AIR DEFENSE

6-12. The SOAR does not possess organic air defense artillery assets. Air defense artillery assets may be available from other sources. As an example, elements of the heavy brigade combat teams and assets of sister Services may be able to provide air defense support. Depending on the theater, Patriot units may provide incidental air defense artillery coverage, as well. Because of its range, the Patriot missile defense system provides coverage well beyond the capabilities of other air defense artillery assets.

6-13. Without proper coordination and integration in the operational environment, the potential exists for conflict between aviation and air defense operations. Air defense artillery units must neutralize hostile aircraft and missiles in the same airspace where friendly aviation units conduct operations. To coordinate each other's efforts, the units observe the following three control statuses of air defense weapons:

- *Weapons hold*: Do not fire except in self-defense or in response to a formal order.
- *Weapons tight*: Engage aircraft only if positively identified as hostile in accordance with published hostile criteria.
- *Weapons free*: Engage aircraft if not positively identified as friendly.

FRATRICIDE

6-14. To reduce the risk of fratricide, units must coordinate with the area air defense commander for routes through borders, locations of friendly forces, or defense zones before mission execution.

This page intentionally left blank.

Chapter 7

Sustainment

INTRODUCTION

7-1. The Army transformation process has resulted in significant changes to the entire theater logistics structure that ARSOF relies on for sustainment. These changes impact virtually every process from theater opening to the distribution of supplies to Soldiers. USASOC reviewed its logistical structures and requirements in concert with the Army's modularity and expeditionary mind-set. USASOC has reorganized to enable an expeditionary logistical force structure to sustain long-duration operations and logisticians dedicated to "sustaining the ARSOF warrior." Detailed information on ARSOF logistics and sustainment is available in FM 3-05.140 and JP 4-0.

CONTINGENCY PLANNING

7-2. In contingency planning, ARSOF units, in conjunction with the SB(SO)(A), prepare a support plan. The support plan identifies support requirements for operation plans and concept plans in a bare-based statement of requirement, down to the user level. The Army Service component command coordinates with in-theater support organizations to fulfill requirements and prepares a support plan identifying support relationships and shortfalls.

7-3. All logistics operations constantly strive to maintain units at a desired level or resource. To maintain the desired level, planners must—

- Maximize the use of existing fixed facilities.
- Limit logistics requirements to mission essentials and acceptable risk.
- Minimize the handling of supplies.
- Concentrate maintenance on returning major end items to service.
- Rely on air lines of communications for rapid resupply.
- Anticipate high attrition of supplies while performing missions in denied areas.
- Identify to the Army Service component command as early as possible those items that require operational floats or other special logistics arrangements.
- Make maximum use of host-nation support, including local and third-country resources.
- Conduct threat assessment.
- Conduct risk assessment.

CRISIS ACTION PLANNING

7-4. Crisis action planning is based on current events and is conducted during time-sensitive situations and emergencies using assigned, attached, or allocated forces and resources. Planners for crisis action planning base their approach on the actual circumstances that exist at the time planning occurs. They follow prescribed procedures that parallel contingency planning, but are more flexible and responsive to changing events and time constraints.

7-5. ARSOF logistics planning must take into consideration the bare-based requirements to support operations. In the early stages of any deployment, ARSOF will normally be required to establish separate intermediate staging bases and eventually expand the number of support bases to meet mission requirements. To maintain the desired level of support, meet operational tempo projections, and provide flexibility, each planner must be able to meet current requirements and to simultaneously plan for future

Chapter 7

ARSOF requirements. Planners must first consider the existing infrastructure in-theater. Using this infrastructure as a baseline, planners then integrate, consolidate, and cross-level resources to maximize logistics support.

LOGISTICS SUPPORT

7-6. Conventional logistics organizations and procedures are normally adequate for SOAR requirements. Standard procedures are in place to handle the few SOF-unique requirements through the TSOC and the Army special operations forces liaison element (ALE). The Theater Sustainment Command provides reception, staging, onward movement, integration, follow-on support, and sustainment of in-theater Army forces. The SOAR has some key differences that impact the type of support required for reception, staging, onward movement, integration, and sustainment. The following conditions occur often enough that they must receive special consideration during logistics planning:

- Forward-deployed units are usually in isolated and austere locations. Distribution is an essential consideration.
- Some special equipment exists; however, most equipment is Army-common, and organic assets can maintain it.

7-7. When the SOAR is normally attached to a SOTF with SF or Ranger logistical organizations, the SF group support battalion or the Ranger regimental support company will be the common-user logistics provider. It will arrange support by coordinating requirements through the SB(SO)(A), ALE, TSOC, and Theater Sustainment Command, and by reachback through the SB(SO)(A) to USASOC and USSOCOM for SOF-unique support. The SF or Ranger logistics organizations are joint- and multinational-capable in that they can accept augmentation of, and employ, common-user logistics assets from ARSOF and other Services and nations. They will then integrate their capabilities into a cohesive plan supporting the commander's operational concept. When ARSOF are assigned to a combined JSOTF, they will provide their organic support packages for Service-specific and common logistics support.

7-8. The TSOC, ALE, SOLE, and logisticians coordinate with the Theater Sustainment Command to develop plans and subsequent orders. They assist in the development of, and implement, directives the commander issues to support the SOAR assigned to the geographic combatant commander. The TSOC and ALE advise the geographic combatant commander on the appropriate command and support relationships for each SOAR mission. The ALE keeps the SB(SO)(A) informed of the status of theater supporting plans and logistical shortfalls.

7-9. The geographic combatant commander supports the SOAR in his area of responsibility. The regiment's logistics planners, with the assistance of the ALE, identify support requirements in the planning phase. They must also identify the logistics shortfalls for inclusion in the geographic combatant commander's risk assessment. If the Theater Sustainment Command cannot support the regiment, it must raise the shortfall to the supported geographic combatant commander for resolution.

SOAR LOGISTICS REQUIREMENTS

7-10. SOAR units require certain logistics support for which the regiment has no organic capability. The regiment's services requirements are as follows:

- Logistics planners must identify and procure tentage for the task force operating in an austere environment. When available, fixed, climate-controlled billeting is optimal for flight management. The regiment must identify billeting requirements.
- The regiment has limited airdrop resupply and equipment maintenance capability. However, it can provide airborne insertion of a forward arming and refueling point and C2 elements. The regiment must identify follow-on airborne requirements. Coordination is through the Theater Sustainment Command or joint task force.
- The regimental S-4 oversees deploying aviation life support system personnel and equipment based on mission variables, mission profile, and duration of the mission. The aviation life support officer ensures preinspections of personal equipment, protective armor, climate kits, and mission-specific equipment. Aviation life support system specialists deploy with limited backup

equipment to support the deploying task force. Additionally, the aviation life support system section provides search and rescue swimmers for overwater operations.
- The regimental task force requests mortuary services, as required. Requests are coordinated through the SOTF and the Theater Sustainment Command.
- The regiment has no food-service capability. Because of mission duration and times, the task force requires rations during 24-hour operations. The regiment must rely on the supported unit to provide food service. The statement of requirements must identify food-service requirements.
- Based on duration of the operation, the regimental task force may require laundry and shower services. When developing the statement of requirements, logistics planners must include water requirements for these services into the total water requirements.
- The regiment has additional water requirements to wash aircraft and flush engines to prevent corrosion during operations in austere environments. Logistics planners must compute these water requirements and identify them in the statement of requirements. Table 7-1 lists the minimum water planning requirements for each type of organic aircraft in remote operations. This water requirement is for manual washing of aircraft and engine flushing on a daily basis. In desert environments, increased water requirements may be required because of the effects of fine sand on the aircraft.

Table 7-1. Water requirements for aircraft washing and engine flushing (gallons)

Type of Aircraft	WATER REQUIREMENTS (GALLONS)		
	Fuselage	Engine	Total
MH/AH-6	25	10	35
MH-60	30	20	50
MH-47	50	21	71

SOAR ORGANIC SUSTAINMENT CAPABILITIES AND SHORTFALLS

7-11. The regiment's collocation of assets with other ARSOF or conventional units reduces external logistics and force health protection distribution problems, and facilitates use of regimental airlift. The following describe organic support capabilities and limitations when not collocated with logistics and sustainment organizations:
- The force deploys with a basic load of meals, ready to eat, for initial sustainment. It has no organic food-service or water storage capability.
- The force deploys with a basic load of Class II supplies for initial sustainment. It has limited document management resources, such as computers, copiers, and shredders.
- Theater pipeline sustainment, joint assets, or in-country sources provide bulk fuel in-theater. During sustained operations, heavy expanded mobility tactical truck fuelers deploy if airlift or sealift is available from the Theater Sustainment Command to provide fuel support at the intermediate staging base or forward staging base. Then the fuelers can establish limited forward arming and refueling points. SOAR units can deploy the equipment by airborne or air landing methods to establish 500-gallon-blivet or 20,000-gallon-bladder forward arming and refueling points, usually in support of a tactical operation. SOAR units do not have the capability to conduct long-term sustainment operations without bulk resupply from theater assets.
- Theater assets must deliver bulk resupply, as the regiment does not have the organic capability to transport large quantities of fuel. Because of the high operational tempo of the unit, the fuel requirement is higher than it is for a similarly sized conventional force. The regiment deploys with a basic load of Class III packaged petroleum, oils, and lubricants for initial sustainment. When appropriate, the regiment requires aerial refueling support for long-range missions. The regiment must identify the refueling requirements as soon as possible.
- Identification of Class IV materiel occurs based on mission requirements in the statement of requirements. Because of the limited space on U.S. Air Force Reserve intertheater airlift allocated for deployment, coordination must occur for pre-positioning and host-nation support.

Chapter 7

- The regiment deploys with a basic load of common and specific Class V supplies. Planners schedule airlift and configure resupply and follow-on ammunition packages for delivery based on the mission. The Theater Sustainment Command or joint task force coordinates ammunition resupply from available sources in-theater. The regimental logistics planners identify common Class V requirements using the statement of requirements. The regiment has a limited capability to transport or store large quantities of Class V supplies and relies on theater transportation and storage.
- Units deploy with Class VI items for initial sustainment (usually a 14- to 30-day supply), when available. Health comfort packets arrive in-theater upon establishment of the logistics system. The SOAR controls weapons systems and replacement aircraft from base stations using limited operational readiness floats. The deployed force requests airframes, weapons systems, and aviation parts through the J-4 to CONUS logistics channels. The regiment's S-4s coordinate with appropriate activities and item managers for immediate release of replacement systems.
- The regiment's flight surgeons develop their deployment load of Class VIII supplies based on the mission variables of mission, enemy, terrain and weather, troops and support available, time available, and civil considerations for initial sustainment. The force then integrates into the joint or theater health service support system for resupply and sustainment.
- Forward support packages deploy with the force. These packages include Class IX air and armament parts and contractor logistics items. The regiment S-4 directs the deployment of the forward support packages based on mission variables and availability of air lines of communication for initial sustainment and follow-on resupply. If air lines of communication are unavailable after deployment for a brief period of time, the forward support section coordinates with the regiment aviation maintenance office and directs additional items to accompany the standard forward support package.
- The ALE coordinates, through the Theater Sustainment Command or SB(SO)(A), Class X supplies for civil-military operations, based on mission variables. Coordination occurs with the joint task force or JSOTF battle staff.

SUSTAINMENT FOR DEVELOPED AND UNDEVELOPED THEATERS

7-12. Once the SOAR is integrated in the Army Service component command logistics system, the regiment will coordinate with the supported theater for logistics resupply. The supported Army Service component command has an ARSOF ALE charged to coordinate logistics for operating in-theater. The ALE is a key element in ensuring logistical requirements meet the regiment's requirements.

7-13. The Army Service component command receives the validated statement of requirements. The Theater Sustainment Command reviews documents (usually during initial and in-progress planning conferences) with the units to determine availability of support and services. The ALE planners coordinate key elements in the theater logistics structure, particularly the Theater Sustainment Command, to support ARSOF. The essential element of support is the establishment of scheduled intertheater and intratheater airlift. Coordination of movement from the home station to the theater is through the SB(SO)(A) to the USASOC G-3.

7-14. Coordination of movement within the theater is through the TSOC or the joint task force joint movement center, with approval for airlift use coming from the geographic combatant commander. This transportation support is the hub of logistics support, since many SOF-unique repair parts, test sets, and associated tools are unavailable in normal theater supply systems. This airlift transports SOF-unique items from origin to the aerial port of debarkation. If the port of debarkation is the destination airfield in the supported theater, the unit (if within range) picks up the repair parts or scheduled intratheater transportation delivers the parts to the destination airfield.

7-15. The deploying SOAR unit must accomplish the following logistical tasks:
- Develop the statement of requirements based on operation plans and mission plans.
- Submit the statement of requirements through operational channels for validation by the TSOC as early as possible but not later than the suspense date.

Sustainment

- Deploy with sufficient required basic loads.
- Schedule additional supplies on later flights as priorities allow.
- Resource the following key personnel to facilitate parts and equipment collection and transfer:
    - Ensure key personnel coordinate with the forward support package manager (deployed) and the supply support activity at Fort Campbell, Kentucky.
    - Coordinate with battalion S-4 representative and production control forward (deployed).
    - Coordinate with battalion S-4 and production control rear at Fort Campbell, Kentucky, who have access to unit technical supply sections able to conduct lateral searches for required items needed forward.
    - Provide regiment S-4 representative in the regiment emergency operations center an information copy of requests (message traffic, fax transmissions) from deployed assets or units upon receipt.

7-16. Upon receipt of a mission or the notification of an impending mission, the regiment HQ begins planning the operation or contingency. Upon notification of authorization to deploy forces, the regiment HQ—

- Implements a 24-hour emergency operations center.
- Provides a forward support package manager for the deploying task force.
- Reviews with the regiment S-3 the statement of requirements from the deploying force and submits these requirements to SB(SO)(A), JSOTF J-4, and the Theater Sustainment Command.
- Provides 24-hour oversight of activities of the supply support activity, aviation life support system, property book officer, organizational clothing and individual equipment, and regimental aviation maintenance officer for aviation-intensive-managed items release.
- Provides property book officer or materiel management for deployed assets.
- Coordinates directly with the designated direct support unit, under "direct liaison authorized."
- Provides a deployment Department of Defense activity address code to the deploying task forces.
- Coordinates (through the ALE) for all local purchases of items not readily available from the Army supply system and SO sources of supply.

FORCE HEALTH PROTECTION SUPPORT

7-17. The SOAR is assigned a flight surgeon, a clinical psychologist, and several special operations combat medics (SOCMs) who are flight-medic-qualified. The regiment does not have physician assistants assigned to its organization. The regiment is dependent upon the theater force health protection assets for Echelon II and above support. Table 7-2 lists the force health protection assets.

Table 7-2. Force health protection personnel authorizations for the SOAR

Unit	Personnel
Special Operations Aviation Regiment	Group Surgeon, Area of Concentration 61N, Major (MAJ)
	Clinical Psychologist, Area of Concentration 73R, MAJ
	Senior Flight Medical Noncommissioned Officer (NCO), MOS 91EW1, SOCM, Sergeant First Class (SFC)
	Flight Medical NCO, MOS 91EW1, SOCM, Sergeant (SGT)
	Flight Medical Specialist, MOS 91EW1, SOCM, Specialist (SPC) (2)
Special Operations Aviation Battalion	Battalion Surgeon, Area of Concentration 62N, Captain (CPT)
Additional Assets	Approximately 15 Flight Medical Specialists, SPC, Distributed Throughout the SOAR
NOTE: Flight medical sergeants and specialists have the special qualifications identifier (SQI) designating flight status.	

FUNDING AND FINANCE SUPPORT

7-18. The finance battalion in-theater provides support for funding and finance service, or the finance battalion supporting the SOAR in garrison (as determined by the task force commander) may provide the service. Funding and finance support includes—
- Providing funds to the ALE or other agents.
- Coordinating resupply of funds, materiel, and services in-theater.
- Coordinating currency exchange with the appropriate embassy or agency.
- Paying local vendors and contracts.

ENGINEER SUPPORT

7-19. The SOAR is dependent upon Army Service component command and theater engineer units for support and sustainment. When available, engineer units conduct a variety of missions, to include the following:
- Engineer reconnaissance teams may assist in reconnaissance missions to locate possible sites for forward arming and refueling points, landing zones, or advanced operating bases.
- Engineers provide current mine threat overlays that may impact ground operations. They clear obstacles and possible booby traps.
- Engineers support countermobility by providing hasty protective row minefield training and by installing obstacles to disrupt, turn, fix, and block enemy forces.
- Engineers construct berms and trenches to protect holding areas and forward arming and refueling points. They help construct wire obstacles around the perimeter. They also help in training camouflage techniques.
- Engineers perform tasks to ensure the continuous sustainment for forward-deployed assets, to include replacement of tactical bridges, support facilities, and area damage control. Tasks also include constructing, maintaining, and repairing combat roads and trails, main supply routes, and lines of communication.
- Engineers provide terrain data in support of air and ground operations. Terrain data help identify possible air corridors, forward arming and refueling point operations, potential landing zones, pickup zones, and terrain that can mask movement.

FORWARD ARMING AND REFUELING POINT OPERATIONS

7-20. The SOAR has an organic airborne forward arming and refueling section that provides Class III (B) and (V) support for operational units. The airborne forward arming and refueling section can rig for airdrop and operating 12-, 16-, or 32-foot Type V platforms with forward arming and refueling point equipment. Forward arming and refueling point personnel can operate MH/AH-6, DAP, MH-60, and MH-47 forward arming and refueling points. They can also operate during joint fixed-wing refueling operations in forward areas. Due to the high volume of fuel required for the MH-47 and MH-60, the tactical airdrop forward arming and refueling points are usually used in support of the MH/AH-6. This aircraft has limited range and lack of in-flight refueling capabilities.

LOGISTICS IN DEVELOPED AND UNDEVELOPED THEATERS

7-21. The uncertainty of today's world presents great challenges for supporting and sustaining ARSOF. These challenges include drug trafficking, natural and man-made disasters, regional conflicts, insurgencies, and terrorist and conventional threats with state-of-the-art weapons. A challenge for ARSOF logistics emerges from the small size of these operations. Small-scale operations will result in smaller, less-developed theaters with little to no dedicated Army Service component command logistics support structure.

7-22. In the early stages of an operation or during crisis response and limited contingency operations, the SB(SO)(A)—in addition to performing reception, staging, onward movement, and integration—may be responsible for establishing the theater-level stockage base and providing logistics support to units deployed forward into their areas of operation. As the theater grows and matures, this sustainment function will transition on order to the sustainment brigade tasked to provide theater distribution and/or to an operational-level sustainment brigade in-theater.

DEVELOPED THEATER

7-23. In a developed or mature theater, a sustainment base sets up within the theater. Pre-positioned war reserve materiel stock and operational project stocks are in place, and foreign nation support agreements exist. Theater Sustainment Command capabilities are normally sufficient to support and sustain ARSOF. In cases where the Theater Sustainment Command is unable to fill ARSOF logistics requests or requirements, the SB(SO)(A) or ALE will exercise their reachback capability to USSOCOM or USASOC to fulfill the requirement.

UNDEVELOPED THEATER

7-24. An undeveloped theater does not have a significant U.S. theater sustainment base. Pre-positioned war reserve materiel stock, in-theater operational project stocks, and foreign nation support agreements are minimal or nonexistent. When an ARSOF unit deploys into an undeveloped theater, it must bring sufficient resources to survive and operate until the Army Service component command establishes a bare-based support system or makes arrangements for host-nation and third-country support. The bare-based support system may function from CONUS, afloat (amphibious shipping or mobile sea bases), or at a third-country support base. The bare-based support system relies heavily on intertheater airlift or sealift for resupply.

CONTRACTING

7-25. Army Tactics, Techniques, and Procedures (ATTP) 4-10, *Operational Contract Support Tactics, Techniques, and Procedures*, and JP 4-0 explain the procedures for obtaining contracting support. Depending on the operational situation and its associated risks, a variety of support functions exist on the battlefield that a contractor can provide or augment. All functions other than those inherently governmental in nature (defined as armed combat, command and control of U.S. military and civilian personnel, and government contracting) or functions covered by host-nation support agreements, may be suitable for contractor support.

7-26. Contracting support on the battlefield is an integral part of the overall process to obtain supplies, services, and construction in support of SOAR operations, and is a critical capability required in an underdeveloped theater. Contracting support can augment existing capabilities, provide expanded sources of supplies and services, bridge gaps in the deployed force structure, leverage assets, and reduce dependence on U.S.-based logistics. Contracting for supplies and services lessens the requirements normally performed by logistics personnel. Contracting personnel should arrive with, or before, the lead ground elements to establish contracting operations. They should depart with, or after, the last ground element to close out the operations. Contracting personnel should establish their operations with, or near, local vendor bases to support deployed forces. Commanders should understand that the contracting officer may need to reside and operate outside a SOTF to be accessible to the local vendor base.

PLANNING CONTRACTING SUPPORT

7-27. Planning permits rapid, coordinated action by staffs and other elements of the command. It also permits the command to respond to rapidly changing situations. Adequate, practical planning is essential to the success of contracting support and is an essential part of the environment of the operational planner and the contract manager. Operational planners at all levels must actively involve contract managers in the planning process to ensure that contracting support is a considered support option and, when used, is responsive to the needs of the command. Success requires advance knowledge of expected support requirements so that a responsive approach can be developed and potential sources identified. Planning for contracting support follows the same process as other planning and is part of both the contingency and crisis action planning processes. Properly included in the planning process, contingency contracting

Chapter 7

personnel locate vendor bases within and near the mission area. They also identify supplies, services, and equipment available from the local economy and advise the commander on how to leverage commercially available support. These measures allow planners to maximize available airlift and sea assets. Central control officers also help commanders avoid basing their plans on false assumptions about the availability or suitability of commercial support.

TYPES OF CONTRACTOR SUPPORT

7-28. Contractors are characterized by the type of support they provide and by the source of their contract authority. Commanders and planners should identify a requirement for a contracted system or capability early on, so all contractors who provide support to the theater and who require transportation can be integrated into the time-phased force and deployment data for timely deployment. Contractor support falls into the three categories discussed below.

Systems Support Contractors

7-29. Systems support contractors logistically support deployed operational forces under prearranged contracts awarded by Service program managers or by military Service component logistic commands. They support specific systems throughout their system's life cycle, including spare parts and maintenance across the range of military operations. These systems include, but are not limited to, weapons systems, C2 infrastructure, and communications systems. For ARSOF, this support is typically arranged by USSOCOM.

External Theater Support Contractors

7-30. External theater support contractors, working pursuant to contracts awarded under the command and procurement authority of the supporting HQ outside the theater, provide support for deployed operational forces. They may be U.S. or third-country businesses and vendors. The contracts are usually prearranged but may be contracts awarded or modified during the mission based on the commanders' needs.

7-31. Services and agencies award the contracts to support U.S. forces in operations worldwide. The services provided by these types of contracts include, but are not limited to, building roads and airfields, dredging, stevedoring, and providing transportation services, billeting and food services, utilities, and decontamination support.

Other Theater Support Contractors

7-32. Theater support contractors support deployed operational forces pursuant to contracts arranged within the mission area, or prearranged contracts through host nation and regional businesses and vendors. Contracting personnel with the deployed force, working under the contracting authority of the theater, the Service component, or the joint task force contracting chief, normally award and administer these contracts. Theater support contractors provide goods, services, and minor construction, usually from the local vendor base, to meet the immediate needs of operational commanders. Immediate contracts involve contracting officers procuring goods, services, and minor construction, either from the local vendor base or from nearby offshore sources, immediately before or during the operation itself.

CONSTRUCTION

7-33. In contracting for construction in contingencies, the Services' agencies designated as Department of Defense construction agents for the peacetime military construction program for specific geographic areas under Department of Defense Directive 4270.5, *Military Construction*, may be used to provide construction contracting in support of military operations. For countries where there is no designated Department of Defense contracting agent, the supported geographic combatant commander usually designates a contract construction agent for support.

Appendix A
SOAR Formats

This appendix contains the formats of an operation order and a mission planning folder specifically designed for SOAR aircrews. Mission-planning aircrews continually coordinate and update these documents throughout the execution of the mission.

SOAR OPERATION ORDER

A-1. The S-3 of the SOAR task force modifies the standard operation order to develop an aviation-specific order (Figure A-1, pages A-1 through A-15). A detailed SOAR operation order is critical to mission success.

(CLASSIFICATION)

1. SITUATION.
   a. Enemy forces.
      (1) Weather (current and forecast).
          (a) Area of operations or objective area.
          (b) Forward staging base.
          (c) En route (ingress and egress).
      (2) Light data.
          (a) Sources of light for night operations.
          (b) Percent of moon illumination.
          (c) Angle of moon during operation.
      (3) Sea data.
          (a) Sea state.
          (b) Water temperature.
      (4) Terrain (area of operations and objective area).
          (a) Key terrain.
          (b) Decisive terrain.
          (c) Avenues of approach (air, land, and sea).
          (d) Cover and concealment.
          (e) Observation and fires.
          (f) Hazards (existing obstacles and minefields).
          (g) Effect on aviation.
          (h) Effect on mission.
          (i) Choke points on route.
      (5) Enemy troops.
          (a) Permissive, uncertain, and hostile environment.

(CLASSIFICATION)

Figure A-1. SOAR operation order format

(CLASSIFICATION)
- (b) En route, landing zone and objective area, forward arming and refueling point sites.
- (c) Identification of forces.
- (d) Locations.
- (e) Strength.
- (f) Morale.
- (g) Capabilities.
- (h) Vulnerabilities.
- (i) Activities (current and future).
- (j) C2.
- (k) Service and support.
- (l) Probable courses of action following mission execution.
- (m) Reaction time from known locations.

b. Friendly forces.
  (1) Higher HQ.
    (a) Command relationship (effective date-time group).
    (b) Mission.
    (c) Intent.
  (2) Ground and assault force.
    (a) Command relationship (effective date-time group).
    (b) Mission.
    (c) Intent.
  (3) Adjacent units.
    (a) Location.
    (b) Mission.
    (c) Airspace coordination.

c. Attachments and detachments.
  (1) Command relationship (effective date-time group).
  (2) Mission.
  (3) Location.

d. Public affairs guidance.
e. Priority intelligence requirements and information requirements.
f. Essential elements of information and essential elements of friendly information.

2. MISSION. Who, what, when, where, and why.
3. EXECUTION.
Intent: State the commander's intent.
  a. Concept of the operation.
    (1) Scheme of maneuver.
      (a) General scheme, mission profile (diagram, chart), and H-hour (if applicable).
      (b) Event- or time-driven.
      (c) Phasing.
      (d) Main effort.

(CLASSIFICATION)

Figure A-1. SOAR operation order format (continued)

(CLASSIFICATION)

- (2) Plan of fire support.
    - (a) General scheme (air, ground, naval).
    - (b) Priority of fires.
    - (c) Target overlay.
    - (d) Types of fires.
    - (e) Preparatory or prehour fires.
    - (f) Fire support coordination measures.
    - (g) Illumination requirements.
    - (h) Suppression of enemy air defenses.
    - (i) Employment of nuclear or chemical fires.
    - (j) Test fire time and location.
    - (k) Actions to prevent fratricide.
- (3) Counterair operations.
    - (a) Assets and plan.
    - (b) Use of aircraft survivability equipment.
    - (c) Passive measures.
- (4) Electronic warfare.
    - (a) Collection and jamming.
    - (b) Types of targets.
    - (c) Priority of jamming.
- (5) Deception.
    - (a) Landing zones, routes.
    - (b) Special movement and landing instructions.
    - (c) Deception target and intent.

b. Tasks to subordinate units.
- (1) Companies, platoons, sections, or teams.
- (2) Aviation unit maintenance.
- (3) HQ.

c. Tasks to combat support units.
- (1) Fire support.
    - (a) Close air support.
    - (b) Close combat attack.
    - (c) Chemical support.
    - (d) Field artillery support (including displacement).
    - (e) Naval gunfire support.
    - (f) Attack helicopters.
    - (g) Special instructions.
    - (h) Fire support overlay and target list.
- (2) Air defense.
    - (a) Command relationship.
    - (b) Specified tasks.

(CLASSIFICATION)

Figure A-1. SOAR operation order format (continued)

Appendix A

(CLASSIFICATION)

- (3) Chemical (including decontamination).
- (4) Electronic warfare.
- (5) Engineering (battle environment preparation).
- d. Intermediate staging base.
  - (1) Marshalling area procedures and control.
  - (2) Time sequence.
    - (a) Show.
    - (b) Concept briefing.
    - (c) Weather decision.
    - (d) Preflight.
    - (e) Aircraft run-ups, equipment checks, communications checks, load time.
  - (3) Route to the area.
  - (4) Fuel requirements.
  - (5) Special equipment required.
  - (6) Contingencies.
    - (a) Minimum number of aircraft.
    - (b) Aircraft abort (cross-loading of personnel and equipment).
    - (c) Weather abort criteria.
    - (d) Ground force command and air mission commander bump plan.
  - (7) Aircraft parking plan.
  - (8) Aircraft load plan on C-5, C-17, and C-130.
  - (9) Security plan.
- e. Forward staging base.
  - (1) Location.
  - (2) Landing.
    - (a) C-5, C-17, C-130.
      - 1. Direction.
      - 2. Time.
      - 3. Offload sequence and position.
      - 4. Aircraft parking and buildup area.
      - 5. Fuel plan.
    - (b) Self-deployed.
  - (3) Aircraft combat load plan.
    - (a) Seat configuration (if any) and number.
    - (b) Straps.
    - (c) Doors opened or closed.
    - (d) Miscellaneous equipment (stored, location).
  - (4) Security requirements.
  - (5) Repositioning for departure.
  - *(6) Hazards.

(CLASSIFICATION)

Figure A-1. SOAR operation order format (continued)

(CLASSIFICATION)

- *(7) Weapons systems loading.
    - (a) Location.
    - (b) Orientation or heading.
    - (c) Safety measures.
- *(8) Takeoff.
    - (a) Time.
    - (b) Heading.
    - (c) Formation.
    - (d) Airspeed.
    - (e) Altitude.
    - (f) Hazards.
    - (g) Aircraft lighting.
    - (h) Aircraft survivability equipment requirements.
    - (i) Fuel required.
    - (j) Weapons status.
- (9) Combat control team instructions.
    - (a) Communications.
    - (b) Signal.
- *(10) Contingencies.
    - (a) Air Force air aborts.
    - (b) Air Force go-around.
    - (c) Aircrew injuries.
    - (d) Bump plan.
        1. Change of lead aircraft because of maintenance.
        2. Change of air mission commander aircraft because of maintenance.
        3. Change of ground commander aircraft because of maintenance.
        4. Change of other aircraft.
        5. Spare (location, running, not running).
- (11) Weather abort criteria.
- (12) Aircraft abort criteria.
    - (a) Minimum number aircraft required to accomplish the mission.
    - (b) Aircraft systems failure criteria.

f. Flight route.
   - (1) Initial approach fix air aborts.
   - (2) Formation.
   - (3) Airspeed and ground speed.
   - (4) Altitudes.
   - (5) Hazards to flight.
   - *(6) Turns in excess of 60 degrees.
   - *(7) Communications signals unique to this portion.

(CLASSIFICATION)

Figure A-1. SOAR operation order format (continued)

Appendix A

(CLASSIFICATION)
- *(8) Air traffic control and combat control team procedures.
- *(9) Aircraft lighting.
- *(10) Checkpoints.
- *(11) Rally points (air and ground).
- *(12) Point of no return.
- *(13) Brief penetration control measures.
    - (a) Identification, friend or foe (IFF), set for penetration.
    - (b) Aircraft survivability equipment and electronic countermeasures (APR-39/44, flares, chaff, ALQ-144, radar, tactical air navigation).
    - (c) Armor panels and plates, forward.
    - (d) Armament systems.
    - (e) Review release point, target, departure, and go-around procedures.
    - (f) Fast rope, extend bars.
    - (g) Fuel transfer off, prior to release point.
    - (h) Warning calls.
    - (i) Aircraft lights, interior and exterior.
    - (j) Forward-looking infrared, lower brightness release point, or short final.
- (14) Release point: time, distance, and heading from release point to objective.
- (15) Egress.
    - (a) Armament systems.
    - (b) Aircraft lights, adjust.
    - (c) IFF.
- (16) Penetration control point procedures.
- (17) Aviation element link-up procedures.
    - (a) Location.
    - (b) Communications.
    - (c) Link-up procedure.
- (18) Aircraft survivability equipment requirements (specify actions along route and especially when cross-forward line of own troops).
- *(19) Weapons status.
    - (a) Hold, fire only in self-defense.
    - (b) Tight, fire only if target identified as enemy.
    - (c) Free, fire at anything except if positively identified as friendly.
- **(20) Contingencies.
    - (a) Downed aircraft.
    - (b) Actions on enemy contact.
    - (c) Communications failure.
    - (d) Lead disorientation.
    - (e) Weather abort.
    - (f) Adjustment to route procedures.
    - (g) Maintenance divert.
    - (h) Casualty evacuation divert.

(CLASSIFICATION)

Figure A-1. SOAR operation order format (continued)

(CLASSIFICATION)
- (i) Formation change.
- (21) Mission abort criteria.
- g. Landing area procedures (landing zone, pickup zone).
  - (1) Location (primary and alternate).
  - (2) Description.
  - *(3) Hazards.
  - *(4) Arrival procedures.
    - (a) Time.
    - (b) Formation.
    - (c) Direction.
    - (d) Airspeed.
    - (e) Air traffic control and combat control team procedures.
    - (f) Aircraft lighting.
    - (g) Doors opened and closed.
    - (h) Door gunner instructions.
    - (i) Signal to execute alternate pickup zone and landing zone.
  - (5) Pickup zone and landing zone marking and control.
  - (6) Aircraft positioning in pickup zone and landing zone.
    - (a) Engines running.
    - (b) Auxiliary power unit only.
    - (c) Complete shutdown.
    - (d) Security in the pickup zone and landing zone.
  - *(7) Load plan (including troop safety considerations).
    - (a) Seat configuration (if any) and number.
    - (b) Strap configuration.
    - (c) Doors opened or closed.
  - *(8) Link-up procedures with supporting aviation elements.
  - *(9) Minimum fuel required to complete the mission from pickup zone and landing zone.
  - (10) Forward arming and refueling point operations (see coordinating instructions).
  - **(11) Contingencies.
    - (a) Lead aircraft change.
    - (b) Air mission commander aircraft change.
    - (c) Task force commander aircraft change.
    - (d) Tactical bump plan.
    - (e) Dispersal plan and rally point.
  - *(12) Time sequencing.
    - (a) Load.
    - (b) Reposition.
    - (c) Takeoff.
    - (d) Time on target or H-hour.
  - *(13) Security requirements.

(CLASSIFICATION)

Figure A-1. SOAR operation order format (continued)

Appendix A

(CLASSIFICATION)

*(14) Weapons status.
(15) Departure procedures (same as forward staging base).
(16) En route procedures (same as flight route).

h. Assault (landing) plan and actions on the objective or target.
   (1) Landing zone and objective location (primary and alternate).
   (2) Time on target and H-hour (rounds on target or wheels down) or critical event that must occur before assault.
   (3) Formation at release point (assault formation).
   (4) Direction of flight.
   *(5) Airspeed and ground speed.
   *(6) Altitude.
   *(7) Aircraft lighting.
   (8) Hazards in the landing zone or objective.
   (9) Landing zone marking and control.
   (10) Air traffic control or combat control team procedures.
   (11) Aircraft touchdown points.
   (12) Type of assault (fast rope, air, land).
   (13) Location of friendly troops in landing zone (reconnaissance and surveillance team, sniper).
   *(14) Aircraft survivability equipment requirements in the landing zone.
   *(15) Air defense status and coordinating instructions.
   *(16) Weapons status (hold, tight, free).
   (17) Fires.
       (a) Door gunners' missions (be specific).
           1. Priorities and sectors.
           2. Shift-fire and hold-fire instructions.
           3. Control measures and actions taken to prevent fratricide.
       (b) Target overlay (artillery, attack helicopter control measures).
       (c) Preplanned fires.
       (d) Rules of engagement.
       (e) Laser safety (troops on the ground).
       (f) DAP, attack helicopter priorities, and type of ammunition.
**(18) Contingencies.
       (a) Downed aircraft on insertion and departure.
       (b) Go-around procedures (direction, signal, altitude, lighting, communication, control measures, and intentions).
       (c) Actions on enemy contact (flight and individual aircraft).
       (d) Communications failure.
       (e) Rally point.
       (f) Friendly killed in action/wounded in action (troops and aircrew members).
       (g) Signal to execute alternate landing zone or objective.
       (h) Aircraft delay and down on landing zone.

(CLASSIFICATION)

Figure A-1. SOAR operation order format (continued)

(CLASSIFICATION)

(19) Departure instructions.
    (a) Authorization to depart.
    (b) Departure plan (when ready, in chalk order, or as a flight).
        NOTE: If departing individually, identify departure sectors, control measures ("Eagle" call), rally point, and safety measures.
    (c) Heading.
    (d) Formation.
    (e) Airspeed and ground speed.
    (f) Aircraft lighting.
    (g) Air traffic control and combat control team procedures.
    (h) Routes (primary and alternate).
(20) Follow-on instructions (laager, prepare for extraction, hold, return to forward staging base).
(21) Safety considerations during the assault and extraction.

i. Departure airfield procedures.
  (1) Location.
  (2) Arrival procedures.
  (3) Parking and teardown.
  (4) Repositioning.
  (5) Marshalling.
  (6) Critical times.
**(7) Contingencies.
    (a) Friendly wounded in action and killed in action.
    (b) Aircraft abort or down.
    (c) Weather abort.
  (8) Security.

j. Coordinating instructions.
  (1) Location and markings of friendly forces.
  (2) Maps and charts.
  (3) Search and rescue plan.
    (a) Brief combat search and rescue zones for ingress, egress, safe area.
    (b) Brief concept of operation, combat search and rescue assets, C2, signal.
    (c) Recovery hospital, secure landing zone and forward staging base.
    (d) Weather or divert plan (alternate recovery instructions).
    (e) Effect on tactical operation.
    (f) Personal locator system coordination.
    (g) Activation or implementation of search and rescue plan.
  (4) Evasion plan of action.
    (a) Escape and evasion route, checkpoints, and procedures.
    (b) Designated area for recovery.
    (c) Recognition and recovery procedures.

(CLASSIFICATION)

Figure A-1. SOAR operation order format (continued)

Appendix A

(CLASSIFICATION)

- (d) Notification procedures.
- (e) Personal locator system and isolated personnel report coordination.
- (f) Destruction of aircraft and sensitive items.

(5) Initial instrument meteorological conditions procedures.
- (a) Altitude to climb to, heading, airspeed, squawk.
- (b) Recovery airfield and instrument capabilities.
- (c) Minimum safe altitude.
- (d) Communications procedures.
- (e) Obstacles, hazards, and highest terrain.
- (f) Effect on mission.
- (g) Forecast weather and freezing level.
- (h) Tactical procedures.
- (i) Approval by higher HQ of initial instrument meteorological conditions recovery plan.

(6) Forward arming and refueling point procedures.
- (a) Location, marking, and marshalling control.
- (b) Airspace management.
- (c) Refueling points.
- (d) Rearm points.
- (e) Safety (weapons, ammunition, personnel).
- (f) Security.
- (g) Dispersal plan (including link-up procedures at rally point).
- (h) Lighting.

(7) Suppression of enemy air defenses.

(8) Aircrew coordinating instructions with assault force in flight (time warnings and navigation).
- (a) Fast-rope procedures, doors open and closed, removal of cargo strap, primary doors.
- (b) Headset and communications coordination.
- (c) Call-out of air control points or release point time, distance, and heading to objective or landing zone.
- (d) Time warnings.
- (e) Confirm landing zone.
- (f) Door fires.
- (g) Clamshell report (report given after all personnel and equipment are clear of the target area).

(9) Mission-oriented protective posture level (including instructions on chemical, biological, radiological, and nuclear defense and decontamination procedures and operational exposure guide).

(10) Rules of engagement.
- (a) Combatants.
- (b) Noncombatants.

(11) Precious-cargo-handling instructions (person or item that is the objective of the mission).
- (a) Handling and control in and out of aircraft.
- (b) Control in the aircraft.
- (c) Doors closed.

(CLASSIFICATION)

Figure A-1. SOAR operation order format (continued)

(CLASSIFICATION)

- (d) Reporting requirements (clamshell report).
- (12) Preaccident plan.
- (13) Inspections.
    - (a) Individual.
    - (b) Aircraft systems and communication.
    - (c) Weapons systems.
- (14) Rehearsals and equipment checks.
    - (a) Loading and offloading procedures.
    - (b) Aerial link-ups.
    - (c) Assault procedures and actions on the objective.
    - (d) Door gunner fire procedures.
    - (e) Communications checks.
    - (f) Aircraft survivability equipment checks.
    - (g) Weapons checks, bore sightings, laser mounts.
    - (h) Precious-cargo handling.
    - (i) Forward arming and refueling point operations.
    - (j) Actions on the objective.
- (15) Air defense warning and air defense weapons control status.
- (16) Individual responsibilities.
    - (a) Flight plan.
    - (b) Manifest.
    - (c) Weather.
    - (d) Sensitive item inventory.
    - (e) Weapons issue.
    - (f) Aviation life support system issue.
    - (g) Mission brief sheet.
    - (h) Sterilization.
    - (i) Fuel identification plate (government credit card).
    - (j) Night vision goggle emergency locator transmitter and radio aircraft key.
- (17) Airspace deconfliction and coordination.
- (18) Precision letdown procedures (Global Positioning System [GPS]) at forward staging base, intermediate staging base, target.
- (19) Antiterrorism measures.
- (20) Backbrief time and location (platoon leader, air mission commander, flight leader, and key personnel; backbrief critical events or concept to commander).
- (21) Debrief time and location.
- (22) Weather decision time.
- (23) Final mission update time and location.

4. SERVICE SUPPORT.
   a. Supply.
      (1) Class I (rations).

(CLASSIFICATION)

Figure A-1. SOAR operation order format (continued)

(CLASSIFICATION)

 (2) Class III (fuel, petroleum, oils, and lubricants).
  (a) Location.
  (b) Type.
  (c) Amount.
  (d) Compatibility of fuel and equipment.
 (3) Class V (ammunition).
  (a) Issue point and procedures.
  (b) Individual authorizations by type.
  (c) Crew-served authorizations by type.
  (d) Aircraft systems authorizations by type.
  (e) Turn-in procedures.
 (4) Class VIII (medical supplies).
 (5) Class IX (aircraft repair parts).

b. Water.
c. Maps.
d. Uniform.
e. Aviation life support system.
f. Special equipment.
 (1) Aircraft survivability equipment.
 (2) Personnel locator system.
 (3) Oxygen.
 (4) Fast-rope insertion and extraction system (FRIES).
 (5) Cargo hook.
 (6) Seats.
g. Storage.
h. Weapons.
 (1) Aircraft systems and aircrew-served weapons.
 (2) Individual weapons.
 (3) Issue instructions.
 (4) Storage and security.
 (5) Test firing and cleaning.
 (6) Maintenance.
 (7) Turn-in procedures.
i. Billeting.
j. Finance.
k. Transportation.
l. Maintenance.
 (1) Location and composition of support.
  (a) Premission.

(CLASSIFICATION)

Figure A-1. SOAR operation order format (continued)

(CLASSIFICATION)

- (b) Preflight.
- (c) Crank.
- (d) Mission.
- (2) Location and composition of ground support equipment.
  - (a) Premission.
  - (b) Preflight.
  - (c) Crank.
  - (d) Mission.
- (3) Support facilities available.
  - (a) Hangar.
  - (b) Shop capabilities.
- (4) Downed aircraft recovery.
  - (a) Recovery officer in charge and team location.
  - (b) Notification on command net frequency of the following information:
    - 1. Type and serial number of downed aircraft.
    - 2. Latitude and longitude of aircraft.
    - 3. Description of recovery site.
    - 4. Brief description of aircraft condition.
    - 5. Area security call sign and frequency.
    - 6. Sensitive items and aircraft configuration.
    - 7. Authority to destroy (if required).
  - (c) Method of recovery.
    - 1. Method of recovery or disposition of aircraft made after consideration of information received and tactical situation.
    - 2. One of three methods of recovery.
- m. Medical evacuation procedures.
  - (1) Responsibilities.
  - (2) Recovery hospital and location.
  - (3) Coordination procedures and communications.
  - (4) Alternate recovery procedure.
  - (5) Collection point at objective area.
  - (6) Launch authority.
- n. Prisoner of war handling and collection point (including civilian detainees, if applicable).
- o. Chemical, biological, radiological, and nuclear equipment and location of decontamination sites.
- p. Services (hygiene, laundry, trash collection).
- q. Discipline, law, and order.
- r. Casualty reporting procedures.

5. COMMAND AND SIGNAL.
  a. Command.

(CLASSIFICATION)

Figure A-1. SOAR operation order format (continued)

Appendix A

(CLASSIFICATION)

(1) Commander and location.
(2) Air mission commander and location.
(3) Assault force commander (second in charge) and location.
(4) Task force commander and location.
(5) Flight lead.
(6) Succession of command (air assault task force commander or ground commander, air component commander, air mission commander, flight or serial commander).
(7) Rear detachment commander.
(8) Battalion or company command post location.
(9) Task force command post location.
(10) Assault force command post location.
(11) Location of command sergeant major or first sergeant.

b. Signal.
  (1) Signal operating instructions period (fill dates).
  (2) Call signs.
  (3) Frequencies or nets (primary and alternate, secure and nonsecure).
      (a) Mission command.
      (b) Task force command.
      (c) Helicopter communications.
      (d) Company internal.
      (e) Adjacent units.
      (f) Fire control.
      (g) Search and rescue.
      (h) Survival, evasion, resistance, and escape, and PLS PRC 112-A-B.
      (i) Medical evacuation.
      (j) Air traffic control and combat control team.
      (k) Airborne warning and control system.
      (l) Vectoring.
      (m) Satellite and high frequency.
  (4) Execution checklist and prowords.
  (5) Signals.
      (a) Lights.
      (b) Visual markers.
      (c) Recognition (day and night).
      (d) Recognition (near and far).
  (6) Challenge and password.
  (7) Running password and number combination.
  (8) Secure communications requirements. Identification of element responsible for keying.
  (9) Transponder.

(CLASSIFICATION)

Figure A-1. SOAR operation order format (continued)

```
                    (CLASSIFICATION)
        (a) Mode requirements (include Transponder Kit IA/C) and codes.
        (b) Antenna requirements.
    (10) Aids to navigation.


  * May be briefed under coordinating instructions.
  ** May be briefed under contingencies.
                    (CLASSIFICATION)
```

Figure A-1. SOAR operation order format (continued)

SOAR MISSION PLANNING FOLDER

A-2. This portion of the appendix provides guidance on preparing the various components of a SOAR mission planning folder. Components of the mission planning folder include the mission tasking letter, the feasibility assessment, the initial assessment, the target intelligence packet for direct action and special reconnaissance missions, the target intelligence packet for FID and unconventional warfare missions, the mission tasking package, the SOF plan of execution, and the plans of execution for infiltrations and exfiltrations.

MISSION TASKING LETTER

A-3. The JFSOCC assigns missions to SOF commanders based on the joint targeting process. Along with the commander who is responsible for executing the mission, the JFSOCC selects a JSOA. Missions may range from a specified task to a broad, continuing mission order. Figure A-2, pages A-15 and A-16, shows a sample format for a mission tasking letter.

```
                        (CLASSIFICATION)
Addressees:
I. REFERENCES.
II. GENERAL.
    A. JFSOCC's concept for employment of the SOAR task force (deployment, employment, and
       sustainment).
    B. JFSOCC's contingency missions assigned to the SOAR task force.
    C. U.S. military strategic objectives.
    D. Variables that complicate the attainment of strategic objectives.
    E. Joint force commander's campaign plan points that accommodate the variables.
    F. Preparation and priorities for unplanned contingencies.
III. JFSOCC CONCEPT OF OPERATIONS.
    A. Location and mission of the JFSOCC.
    B. SOAR task force mission.
IV. SPECIFIC MISSION GUIDANCE.
    A. Force operating locations.
    B. Readiness standards (deployment time and time to commence operations expressed in N+XX).
    C. Specified operations tasks.
                        (CLASSIFICATION)
```

Figure A-2. Mission tasking letter format

Appendix A

(CLASSIFICATION)

D. Specified planning and preparation tasks.

E. Specified area and mission orientation.

F. Actions to prevent fratricide.

V. SUSTAINMENT.

VI. COMMAND AND CONTROL.

VII. OTHER RESPONSIBILITIES.

(CLASSIFICATION)

Figure A-2. Mission tasking letter format (continued)

FEASIBILITY ASSESSMENT

A-4. The feasibility assessment (Figure A-3, pages A-16 and A-17) is a basic target analysis that provides an initial determination of the viability of a proposed target for SOF employment. Normally, the feasibility assessment is associated with the SOF ground element. When a SOAR is given a unilateral mission to conduct a direct action or a special reconnaissance mission, the feasibility assessment should be used for target analysis. It should include an initial assessment.

(CLASSIFICATION)

SECTION I - MISSION DESCRIPTION

A. Target Identification Data (Basic encyclopedia number, category code, geographic coordinates, universal transverse mercator coordinates, and map sheets).

B. Mission Statement and Commander's Guidance.

SECTION II - COMMANDER'S ASSESSMENT

A. Target Feasibility.

B. Probability of Mission Success.

C. Recommendation.

SECTION III - ASSUMPTIONS

SECTION IV - FACTORS AFFECTING COURSES OF ACTION

A. Characteristics of Joint Special Operations Area.

   1. Weather.

   2. Terrain.

   3. Other factors.

B. Friendly Situation.

C. Enemy Situation.

   1. Composition.

   2. Disposition.

   3. Strength.

      a. Committed forces.

      b. Location of reinforcements and estimated reaction times.

      c. Chemical, biological, radiological, and nuclear capabilities.

   4. Significant enemy activity, intelligence, and counterintelligence capabilities.

(CLASSIFICATION)

Figure A-3. Feasibility assessment format

SOAR Formats

```
(CLASSIFICATION)
    5. Peculiarities and weaknesses.
    6. Vulnerability to deception.
    7. Enemy capabilities.
        a. Defensive.
        b. Offensive.
        c. Intelligence and counterintelligence.
    8. Reaction and reinforcement.
    9. Security on target.
SECTION V - COURSES OF ACTION
    A. Identification of Courses of Action.
    B. Analysis of Courses of Action.
    C. Comparison of Courses of Action.
        1. Advantages.
        2. Disadvantages.
        3. Risks.
    D. Recommended Course of Action.
SECTION VI - INTELLIGENCE REQUIREMENTS
SECTION VII - SPECIAL REQUIREMENTS
    A. Personnel.
    B. Logistics.
    C. Other.
(CLASSIFICATION)
```

Figure A-3. Feasibility assessment format (continued)

INITIAL ASSESSMENT

A-5. The initial assessment (Figure A-4, pages A-17 through A-19) provides a basic determination of the viability of the infiltration and exfiltration portions of a proposed SOF mission. If the mission is to support another Army or Navy SOF mission, the initial assessment becomes a part of the feasibility assessment as needed to help establish the overall viability of the mission. The initial assessment goes to the mission planning agent for review and approval before it goes to the JFSOCC.

```
(CLASSIFICATION)
SECTION I - MISSION DESCRIPTION (Basic encyclopedia number, category code, geographic coordinates,
    universal transverse mercator coordinates, and map sheets.)
SECTION II - ASSUMPTIONS
SECTION III - MISSION DATA
    A. Launch bases, intermediate staging bases, and recovery bases.
    B. Landing zones, drop zones, seaward launch points, beach landing sites, recovery zones, and seaward
        recovery points.
    C. Abort and emergency divert bases.
    D. Air-refueling tracks and forward arming and refueling points.
(CLASSIFICATION)
```

Figure A-4. Initial assessment format

Appendix A

(CLASSIFICATION)
   E. Flight and seaward approach routes.
      1. Ingress.
      2. Egress.
      3. Orbiting and holding.
   F. Range factors.
   G. Time factors.
   H. Route factors.
   I. Refueling factors.
   J. Aircrew factors.
   K. Weather.
SECTION IV - MISSION ASSESSMENT
   A. Threat.
      1. Air defenses.
      2. Deception of air defenses.
      3. Surface and subsurface waters.
   B. Probability of team infiltration.
   C. Probability of team resupply.
   D. Probability of team exfiltration.
   E. Overall probability of mission success.
   F. Other factors.
SECTION V - LIMITING FACTORS
   A. Intelligence.
   B. Weather.
   C. Terrain and hydrography.
   D. Equipment.
   E. Monitors.
   F. Tactics.
   G. Logistics.
   H. Personnel.
   I. Training.
   J. Supporting forces.
   K. Rules of engagement, fratricide prevention, and legal issues.
SECTION VI - SUPPORTING DATA
   A. Photography and imagery requested.
   B. Intelligence information requested.
SECTION VII - INITIAL ASSESSMENT BOARD
   A. Composition.
   B. Recommendation.
(CLASSIFICATION)

Figure A-4. Initial assessment format (continued)

(CLASSIFICATION)

SECTION VIII - SOA, SURFACE SHIP, AND SUBMARINE REQUIREMENTS FROM AIR FORCE SOF, ARSOF, AND NAVY SOF TO CONDUCT INITIAL ASSESSMENT

A. Target coordinates.

B. Maximum and minimum distances of landing zones, drop zones, seaward launch points, beach landing sites, recovery zones, and seaward recovery points from target.

C. Time frame in operation plan and concept plan scenario (Pre-D-Day or D+XX).

D. Desired launch and recovery bases.

E. Type of delivery and recovery required (airdrop, airland, fast rope, SEAL delivery vehicle, or combat rubber raiding craft) and preferred platform.

F. Number of personnel to be transferred and approximate weight per person.

G. Approximate size and weight of additional equipment.

H. Assumptions made during supported commander's feasibility assessment.

NOTE: Although not always readily available, information on hand is normally sufficient to conduct the initial assessment. An effort should be made, however, to obtain and include in the initial assessment all the information in the format sample.

(CLASSIFICATION)

Figure A-4. Initial assessment format (continued)

TARGET INTELLIGENCE PACKAGE FOR DIRECT ACTION AND SPECIAL RECONNAISSANCE MISSIONS

A-6. Figure A-5, pages A-19 and A-20, provides a sample format of a SOF target intelligence package (TIP) for direct action and special reconnaissance missions.

(CLASSIFICATION)

SECTION I - TARGET IDENTIFICATION AND DESCRIPTION

A. Target identification data.

B. Description and significance.

C. Detailed target description.

D. Target vulnerability assessment.

SECTION II - NATURAL ENVIRONMENT

A. Geographic data (including terrain and hazards to movement).

B. Meteorological data (climatology overview and tables and illumination data).

C. Hydrographic data (coasts, waterways, lakes, and luminescence).

SECTION III - THREAT

A. Ground forces (including border guards).

B. Paramilitary and indigenous forces (including intelligence and security and police services).

C. Naval forces (including coast guard and maritime border guard).

D. Air forces.

E. Air defense forces (including radars, passive detectors, and communications systems).

F. Electronic order of battle.

(CLASSIFICATION)

Figure A-5. TIP format for direct action and special reconnaissance missions

Appendix A

(CLASSIFICATION)

   G. Space-based assets.

   H. Other.

SECTION IV - DEMOGRAPHICS AND CULTURAL FEATURES

   A. Area population characteristics.

   B. Languages, dialects, and ethnic composition.

   C. Social conditions.

   D. Religious factors.

   E. Political characteristics.

   F. Economic conditions.

   G. Miscellaneous (currency, holidays, dress, and customs).

SECTION V - LINES OF COMMUNICATION AND INFORMATION SYSTEMS

   A. Airfields.

   B. Railways.

   C. Roadways.

   D. Waterways.

   E. Ports.

   F. Petroleum, oils, and lubricants.

   G. Power grid.

   H. Telecommunications and media (print, radio, and television).

SECTION VI - INFILTRATION AND EXFILTRATION (Potential landing zones, drop zones, beach landing sites, and helicopter landing zones.)

   A. Potential zones.

   B. Choke points between insertion points and objective.

SECTION VII - SURVIVAL, EVASION, RESISTANCE, ESCAPE, AND RECOVERY DATA

   A. Survival, evasion, resistance, escape, and recovery (SERER) and safe areas.

   B. Survival data.

SECTION VIII - UNIQUE INTELLIGENCE (Mission-specific requirements not covered above.)

SECTION IX - INTELLIGENCE SHORTFALLS

Annex A. Bibliography.

Annex B. Glossary.

Annex C. Imagery.

Annex D. Maps and charts.

Annex E. Sensitive compartmented information (if applicable).

(CLASSIFICATION)

Figure A-5. TIP format for direct action and special reconnaissance missions (continued)

TARGET INTELLIGENCE PACKAGE FOR FOREIGN INTERNAL DEFENSE AND UNCONVENTIONAL WARFARE MISSIONS

A-7. Figure A-6, pages A-21 and A-22, shows a sample format of a TIP for FID and unconventional warfare missions.

(CLASSIFICATION)

SECTION I - OBJECTIVE AREA IDENTIFICATION AND DESCRIPTION
  A. Objective area identification data.
  B. Description and significance.

SECTION II - NATURAL ENVIRONMENT
  A. Meteorological data (illumination data and climatology overview and tables).
  B. Hydrographic data (coasts, waterways, lakes, and luminescence).
  C. Water sources (color-coded overlay).
  D. Flora and fauna (plants and animals of tactical importance).

SECTION III - THREAT
  A. Objective country (enemy order of battle).
  B. Opposition and resistance forces.

SECTION IV - DEMOGRAPHIC, CULTURAL, POLITICAL, AND SOCIAL FEATURES (Essential elements of information must be answered for the objective country and for opposition and resistance forces.)
  A. Area population characteristics (including resistance potential).
  B. Languages, dialects, and ethnic composition.
  C. Social conditions.
  D. Religious factors.
  E. Political characteristics.
  F. Available labor force.
  G. Customs (society, weapons, religion, culture, and mores).
  H. Medical capabilities.
  I. Health and sanitation conditions.
  J. Economic conditions and influences.
  K. Currency and legal holidays.

SECTION V - LINES OF COMMUNICATION, INFORMATION SYSTEMS, AND LOGISTICS
  A. Airfields.
  B. Railways.
  C. Roadways.
  D. Waterways.
  E. Ports.
  F. Petroleum, oils, and lubricants.
  G. Power grid.
  H. Telecommunications and media (print, radio, and television).
  I. Exploitable civilian transportation.
  J. Primary modes of transportation.
  K. U.S.-provided materials and services.
  L. Stockpiles.
  M. War-sustaining industries.
  N. War-sustaining resupply.
  O. Movement control centers.

(CLASSIFICATION)

Figure A-6. TIP format for foreign internal defense and unconventional warfare missions

Appendix A

(CLASSIFICATION)

SECTION VI - INFILTRATION AND EXFILTRATION (Potential landing zones, drop zones, beach landing sites, and helicopter landing zones.)

    A. Potential zones.

    B. Choke points between insertion points and the objective country. (Essential elements of information must be answered for the objective country and for opposition and resistance forces.)

SECTION VII - FID AND MILITARY ASSISTANCE (Essential elements of information must be answered for the objective country and for opposition and resistance forces.)

    A. Military assistance provided.

    B. Foreign personnel (noncombatants).

    C. Foreign military materiel.

    D. Deployments of foreign personnel and equipment.

    E. Foreign contractor services and construction.

    F. U.S. support.

SECTION VIII - SURVIVAL, EVASION, RESISTANCE, ESCAPE, AND RECOVERY DATA

    A. Survival, evasion, resistance, escape, and recovery; safe areas; and designated areas for recovery.

    B. Survival data.

SECTION IX - UNIQUE INTELLIGENCE (Mission-specific requirements not covered above.)

SECTION X - INTELLIGENCE SHORTFALLS

Annex A. Bibliography.

Annex B. Glossary.

Annex C. Imagery.

Annex D. Maps and charts.

Annex E. Sensitive compartmented information (if applicable).

(CLASSIFICATION)

Figure A-6. TIP format for foreign internal defense and unconventional warfare missions (continued)

MISSION TASKING PACKAGE

A-8. Each mission tasking package consists of a mission tasking letter and the transmittal documents. Figure A-7, pages A-22 and A-23, shows the items included in the mission tasking package.

(CLASSIFICATION)

SECTION I - MISSION TASKING LETTER AND TRANSMITTAL DOCUMENTS

    A. JFSOC tasking.

    B. Subordinate tasking from the JFSOCC.

    C. Coordinating instructions.

    D. Direct liaison authorized (yes or no).

SECTION II - TARGET IDENTIFICATION DATA

    A. Name.

    B. Basic encyclopedia number.

(CLASSIFICATION)

Figure A-7. Mission tasking package format

SOAR Formats

```
                           (CLASSIFICATION)
  C. Mission number (if applicable).
  D. Mission tasks.
  E. Functional classification code.
  F. Country.
  G. JSOA coordinates (geographic reference or universal transverse mercator grid).
  H. Geographic coordinates (geographic reference or universal transverse mercator grid).
    I. General description and target significance.
  SECTION III - COMBATANT COMMANDER MISSION GUIDANCE (Combatant commander's mission
  statement and objectives.)
    A. Mission statement.
    B. Specific targeting objective.
    C. Commander's guidance.
    D. C2.
  SECTION IV - RECORD OF CHANGES
  SECTION V - RECORD OF DISTRIBUTION
                           (CLASSIFICATION)
```

Figure A-7. Mission tasking package format (continued)

SPECIAL OPERATIONS FORCES PLAN OF EXECUTION

A-9. The plan of execution is a detailed plan that shows how the SOF mission will be carried out. This plan, along with the plan of exeuction for infiltration and exfiltration and mission rehearsals, is the result of the targeting and mission-planning process. It also describes the supporting infiltration and exfiltration plan developed by the supporting organization. Figure A-8, pages A-23 through A-25, shows a sample SOF plan of execution format.

```
                           (CLASSIFICATION)
  Issuing HQ:
  Place:
  Day, Month, Year, Hour:
  Commander's or Mission Planning Agent's Estimate of the Situation:
  References (Maps, Charts, and Other Pertinent Documents):
  SECTION I - MISSION DESCRIPTION (Basic encyclopedia number, category code, geographic coordinates,
    universal transverse mercator coordinates, and map sheets.)
  SECTION II - THE SITUATION AND COURSES OF ACTION
    A. Considerations affecting the courses of action.
      1. Characteristics of the JSOA.
        a. Military geography.
          (1) Topography.
          (2) Hydrography and luminescence data.
          (3) Climate, weather, and illumination data.
        b. Transportation.
                           (CLASSIFICATION)
```

Figure A-8. SOF plan of execution format

(CLASSIFICATION)

   c. Telecommunications.
   d. Politics.
   e. Economics.
   f. Sociology.
   g. Science and technology.
2. Relative combat power.
   a. Enemy.
      (1) Strength.
      (2) Composition.
      (3) Location and disposition.
      (4) Reinforcements.
      (5) Logistics.
      (6) Time and space factors.
      (7) Combat efficiency.
   b. Friendly forces.
      (1) Strength.
      (2) Composition.
      (3) Location and disposition.
      (4) Reinforcements.
      (5) Friendly force assistance.
      (6) Logistics.
      (7) Time and space factors.
      (8) Combat efficiency.
3. Assumptions.
B. Analysis of enemy capabilities.
C. Comparison of friendly courses of action.
   1. Statement of courses of action.
   2. Assessment of the probability of success.
   3. Comparison of courses of action.
D. Decision (recommended course of action) and mission profile.
   1. Method and location of infiltration.
   2. Movement to target area.
   3. Actions at the objective.
   4. Movement to the objective and the method of exfiltration.

SECTION III - SUPPORTING PLANS
A. Overall schedule.
   1. Preparation.
   2. Rehearsal.
   3. Rendezvous.
   4. Transit.

(CLASSIFICATION)

Figure A-8. SOF plan of execution format (continued)

```
                        (CLASSIFICATION)
    5. Execution.
    6. Recovery.
  B. Logistics.
  C. Communications and electronics procedures and equipment operating instructions.
  D. Deception.
  E. Indigenous force support.
  F. Time and distance charts.
  G. Deployment.
  H. Weapons.
  I. Target recoverability.
  J. Resupply.
  K. Exfiltration.
  L. Survival, evasion, resistance, escape, and recovery.
  M. Command relationships.
  N. Military Information Support operations and Civil Affairs.
SECTION IV - LIMITING FACTORS
  A. Intelligence.
  B. Weather.
  C. Equipment.
  D. Tactics.
  E. Logistics.
  F. Personnel.
  G. Training.
  H. Supporting Forces.
  I. C2 and communications.
  J. Law of war, rules of engagement, and U.S. law and legal issues.
  K. Other factors.

(Signed)
Commander

ANNEXES (As required. List letter and title.)
DISTRIBUTION (According to policies and procedures of the issuing HQ and at the direction of the JFSOCC.)
                        (CLASSIFICATION)
```

Figure A-8. SOF plan of execution format (continued)

PLAN OF EXECUTION FOR INFILTRATION AND EXFILTRATION

A-10. The plan of execution for infiltration and exfiltration is a detailed plan that shows exactly how the SOA will execute its assigned mission. This plan, along with the supported element's plan of execution and mission rehearsals, is the result of the targeting and mission-planning process. It includes fixed- and rotary-wing aircraft, surface ships, and submarines. Figure A-9, pages A-26 through A-28, shows a sample format of a plan of execution for infiltration and exfiltration.

Appendix A

(CLASSIFICATION)

SECTION I - MISSION
  A. Target identification data.
  B. Mission statement.

SECTION II - MISSION SUMMARY
  A. Mission tasking.
  B. Objective area.
  C. General concept.
  D. Summary of limiting factors.
  E. Probability of mission success.

SECTION III - ASSUMPTIONS

SECTION IV - THREAT ASSESSMENT

SECTION V - NAVIGATION AND OVERALL MISSION PORTRAYAL (This section represents the entire infiltration and exfiltration route from launch to recovery on a suitable scale chart. It shows information the planning cell deemed necessary to portray the mission. Items listed, however, are not all-inclusive.)
  A. Launch bases.
  B. Intermediate staging bases.
  C. Landing zones, drop zones, beach landing sites, recovery zones, seaward launch points, and seaward recovery points.
  D. Recovery bases.
  E. Abort and emergency diversion.
  F. Air-refueling tracks and forward arming and refueling points.
  G. Routes.
    1. Ingress.
    2. Egress.
    3. Orbiting and holding.
    4. Safe passage procedures.
    5. Strip charts, navigation logs, GPS receivers, and other aids (as required).

SECTION VI - SUPPORTING PLANS
  A. Overall schedule of events.
  B. Prelaunch requirements.
    1. Updates to order of battle.
    2. Essential elements of information.
    3. Problem areas and key factors.
  C. Infiltration and exfiltration platform factors and logistical considerations.
  D. C2 and communications.
    1. Security preparations.
    2. Departure procedures (overt or deception procedures).
    3. Communication equipment requirements.
      a. Infiltration and exfiltration platforms.
      b. Supported special operations ground component.

(CLASSIFICATION)

Figure A-9. Infiltration and exfiltration plan of execution format

(CLASSIFICATION)

    4. Specialized operational procedures and techniques.

    5. Radio silence areas.

    6. GO, NO-GO point.

    7. Publication of joint signal operating instructions for air mission.

    8. Deception.

E. Emergency.

    1. Engine-out capabilities.

    2. Weather.

    3. Faulty intelligence.

    4. Infiltration and exfiltration platform abort procedures.

        a. Late-departure procedures.

        b. Maintenance problems.

        c. Battle damage.

        d. Aircraft destruction plan.

        e. Bump plan.

    5. Drop or other fuel-related malfunctions.

    6. Lost communications procedures.

    7. Mission abort procedures.

F. Evasion plan of action.

    1. Aircrew responsibilities.

    2. Immediate actions upon sinking, ditching, or bailing out.

    3. Evasion movement.

    4. Safe area intelligence descriptions.

    5. Selected area for evasion.

    6. Evasion team communications.

    7. Search and rescue contact procedures.

SECTION VII - LIMITING FACTORS

A. Intelligence.

B. Weather.

C. Equipment.

D. Munitions.

E. Tactics.

F. Logistics (including sustainment).

G. Personnel.

H. Training.

I. Supporting forces.

J. Rules of engagement and legal issues.

(CLASSIFICATION)

Figure A-9. Infiltration and exfiltration plan of execution format (continued)

Appendix A

(CLASSIFICATION)

SECTION VIII - SOA, SURFACE SHIP, AND SUBMARINE REQUIREMENTS FROM AIR FORCE SOF, ARSOF, AND NAVY SOF TO CONDUCT INITIAL ASSESSMENTS

A. Target coordinates.

B. Maximum and minimum distances of landing zones, drop zones, seaward launch points, beach landing sites, recovery zones, and seaward recovery points from the target.

C. Time frame in operation plan and concept plan scenario (Pre-D-Day or D+XX).

D. Desired launch and recovery bases.

E. Type of delivery and recovery required (airdrop, airland, fast rope, SEAL delivery vehicle, combat rigid rubber raiding craft, and platform preferred).

F. Number of personnel to be transferred and approximate weight per person.

G. Approximate size and weight of additional equipment.

H. Assumptions made during supported unit's feasibility assessment or plan of execution.

I. Desired time over target.

J. Resupply and exfiltration requirements.

NOTE: Although not always readily available, information on hand is normally sufficient to conduct the initial assessment. An effort should be made, however, to obtain and include in the plan of execution all the information in this sample format.

(CLASSIFICATION)

Figure A-9. Infiltration and exfiltration plan of execution format (continued)

SPECIAL OPERATIONS AVIATION CALL-FOR-FIRE FORMAT

A-11. To employ SOA attack helicopters for fire support, the following 5-line call-for-fire format will be used (Figure A-10, pages A-28 and A-29).

---

1. Observer ID/Warning Order

    " (Aircraft Call Sign), this is (Observer Call Sign), fire mission over."

    - **Observer ID:** This element of the call for fire tells the fire direction center who is calling for fire.
    - **Warning Order:** Clears the net for the fire mission and tells the aircraft the *type of mission* that will be used.
      Type of mission: Fire mission (standard), illumination, Hellfire, dual targets, flechette, or recon (recce).

2. Friendly Location/Mark

    "My position (Target reference point, grid, building) marked by (Strobe, beacon, VS-17 panel)."

    - **Friendly location:** This will be the observer's location. Any other friendly elements or the closest friendlies will be sent in the remarks section. All friendly positions should be marked and announced to the pilot.

3. Target Location

    "Target Location (Bearing [magnetic] and range [meters], target reference point, grid, building)."

    - **Target Location:** Bearing will be announced in magnetic and will be expressed as single digits. Example: 130 degrees will be announced as "ONE THREE ZERO." Range will be expressed as it would normally be said. Example: 200 meters will be announced TWO HUNDRED.

Figure A-10. Special operations aviation call-for-fire format

SOAR Formats

4. Target Description/Mark

"(Target Description), marked by (IR sparkle, tracer, and so on)."

- **Target Description**: What it is, what it is doing, number of elements, size and shape, and degree of protection. Target description should be quick and concise. If the pilot does not acknowledge the target, then you will give a talk on after remarks.

5. Remarks, as required (danger close clearance, restrictions, fire support coordination measures, at my command, and so on). See notes below.

"Over."

NOTE: Danger close fire missions require a statement in line 5: "Cleared Danger Close," the ground force commander's initials, and the location of the closest friendly troops with respect to the target. A readback of "Danger Close" and confirmed location of closest friendly troops by attacking platform is required.

Line five: Is used as required to provide as much additional information as needed to build situational awareness in the conduct of the call for fire and execution of the fire mission.

Clearance: Transmission of the 5-line call-for-fire brief is clearance to fire and equals consent from the ground force commander (unless danger close). For "Danger Close Fire," the observer/commander must accept responsibility for increased risk. State "Cleared Danger Close" in line 5 along with ground force commander's initials. This clearance may be preplanned.

Postmission: Immediately following execution of fire mission, give battle damage assessment (BDA), corrections, end of mission, or new target, as required.

Example:

<u>Observer/Warning Order</u>:

Line 1. **"B41, this is V99, fire mission over."**
   "V99, send your fire mission."

Lines 2–4. From V99—(*Friendly Location/Mark*) **"My position east side of building # eight, marked by IR strobe."** (*Target Location*) **Three-three-zero degrees, one hundred sixty meters.** (*Target Description/Mark*) **Target is four enemy pax with small arms west side of building # one, marked by IR sparkle.**

NOTE: Lines 2–4 are sent in conjunction with each other.

<u>Remarks</u>:

Line 5. Reply B41—**"V99, I have your position east side building # eight, target three-three-zero degrees, one hundred sixty meters, four pax west side of building # one. Tally target."**

NOTE: Rotary-wing call for fire is a request for fire and does not need clearance from joint terminal attack controller/observer. If additional control measures need to be emplaced, the phrase "*At My Command*" may be placed at the end of Line 4. The pilot should acknowledge the control measure in the remarks.

<u>Post-mission observer action</u>:

From Y99—**"B41, end of mission, target destroyed."**

Reply B41—**"B41 copies, end of mission, target destroyed."**

Figure A-10. Special operations aviation call-for-fire format (continued)

This page intentionally left blank.

Appendix B
Aircraft Capabilities

This appendix lists all aircraft in the SOAR inventory. It defines the most advanced technology in rotary-wing aircraft. Because of continual advancements in technology, the capabilities of all aircraft and systems are continually changing. This appendix provides a capabilities matrix that lists all SOAR aircraft. The matrix serves as an easy and rapid reference for mission planners.

MH-6M HELICOPTER

B-1. The primary mission of the MH-6M helicopter is to conduct overt and covert infiltration, exfiltration, and combat assaults over a wide variety of terrain and environmental conditions. The MH-6M also performs C2 and reconnaissance missions.

DESCRIPTION

B-2. The MH-6M is a light assault helicopter (Figure B-1). It is a single-engine, light utility helicopter modified to transport up to six combat troops and their equipment externally. Its small size allows for rapid deployability in C-130, C-17, and C-5 transport aircraft. Aircraft modifications and aircrew training allow for extremely rapid upload and download times.

Figure B-1. MH-6M helicopter

Appendix B

AIRCRAFT SURVIVABILITY EQUIPMENT

B-3. The MH-6M has no standard aircraft survivability equipment other than the APR-39 Radar Warning Receiver System. This passive omnidirectional warning set detects and identifies hostile search and acquisition and fire control radar. It provides audio and visual alerts to the flight crew.

STANDARD MISSION EQUIPMENT

B-4. Some aircraft have forward-looking infrared, a passive imaging system that detects long-wavelength radiant infrared energy emitted, naturally or artificially, by any object in daylight or darkness, and provides an infrared image of terrain features and ground or airborne objects of interest. A standard computer equipped with a Personal Computer Memory Card International Association (PCMCIA) card reader and Video LAN Client (VLC) Media Player can play back recorded images.

B-5. The MH-6M can have a combination of internal or external auxiliary fuel tanks installed as required for the mission. The internal tank, the Goliath, is mounted in the passenger area and provides 62 additional gallons of fuel which adds approximately 80 minutes of flight time. The external tank, the Improved Conformal Equipment auxiliary fuel tank, is mounted under the pod and provides 31 additional gallons of fuel which adds approximately 40 minutes of flight time. Both auxiliary tanks are ballistic-tolerant up to .50 caliber.

ARMAMENT

B-6. The MH-6M has no standard armament.

SPECIAL MISSION EQUIPMENT

B-7. Personnel can rapidly configure the aircraft for fast-rope operations or caving ladder extraction.

TRANSPORTABILITY OF MH-6M AIRCRAFT

B-8. A C-130 can carry 3 MH-6Ms, a C-17 can carry 9, and a C-5 can carry 21. In each case, tactical uploading and downloading of the aircraft can take place in an extremely short time.

PLANNING CONSIDERATIONS

B-9. The following paragraphs discuss considerations that must be taken into account when planning to use MH-6M aircraft in a mission.

Weather Minimums

B-10. A minimum 500-foot ceiling and a 2-mile visibility capability must exist for day and night flying over all types of terrain. The unit commander may reduce weather minimums on a case-by-case basis.

B-11. A visible horizon must exist in two of the four horizontal quadrants at all times. All MH-6M missions must take place under visual meteorological conditions. Instrument flight rule flights are unauthorized.

Winds

B-12. The maximum wind allowed to start the aircraft is 40 knots, with a 20-knot gust spread.

Flight Altitudes

B-13. For training missions, the minimum en route altitude for routes not reconnoitered is 300 feet above ground level. The minimum overwater altitude is 50 feet. For operational missions, the minimum en route altitude is dependent upon mission, enemy, terrain and weather, troops and support available, time available, and civil considerations.

Aircraft Capabilities

Landing Areas

B-14. The MH-6M is capable of landing on any structure that allows clearance for the rotor systems (30 feet) and meets stress requirements. Single-aircraft confined landing areas require a minimum size of 50 feet by 50 feet.

Shipboard Operations

B-15. The MH-6M can operate day and night from any ship with at least a one-spot helicopter-landing capability.

Aircrew Composition

B-16. The normal aircrew for most training exercises and operational or contingency missions consists of a pilot and a copilot. All overwater flights require a pilot and copilot current and qualified in overwater flight. All aircrews can conduct night vision goggle infiltration and exfiltration, stabilized body operations, FRIES, and aerial suppression operations to urban, mountainous, desert, and jungle objectives, as well as to ships and offshore drilling platforms. Aircrews have training in long-range precision navigation and formation flight over land and water to arrive at objectives at a prearranged time (plus or minus 30 seconds).

Aircraft Capabilities

B-17. Table B-1 lists the capabilities of the MH-6M aircraft. Figures B-2 and B-3, page B-4, illustrate specific dimensions of the aircraft.

Table B-1. MH-6M aircraft capabilities

Aircraft Weight	
Basic weight	2,300 pounds
Mission weight	3,500 pounds
Maximum gross weight	4,700 pounds
Aircraft Dimensions	
Length, blades folded	22 feet 6 inches
Length, blades unfolded	32 feet 1 inch
Width, blades folded	6 feet 5 inches
Width, blades unfolded	27 feet 4 inches
Height	8 feet 11 inches
Diameter of main rotor	27 feet 4 inches
Aircraft turning radius	36 feet 9 inches

Range and Endurance at 240 Pounds per Hour				
Fuel Tank	Endurance (Hours + Minutes)	Aircrew	Passenger	Fuel Range (Nautical Miles)
Main	1+00	2	5	80
Main plus improved conformal equipment tank	1+40	2	4	135
Main plus Goliath tank	2+20	2	3	190
Main plus Goliath and improved conformal equipment tanks	3+00	2	2	240

Airspeed. The cruise airspeed for the MH-6M helicopter is 80 knots.

Appendix B

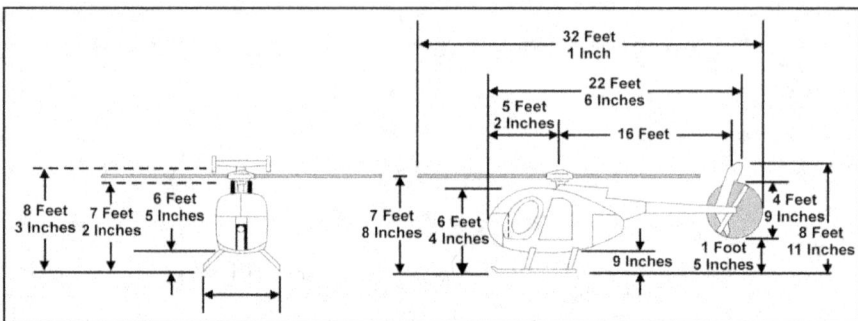

Figure B-2. MH-6M and AH-6M aircraft dimensions

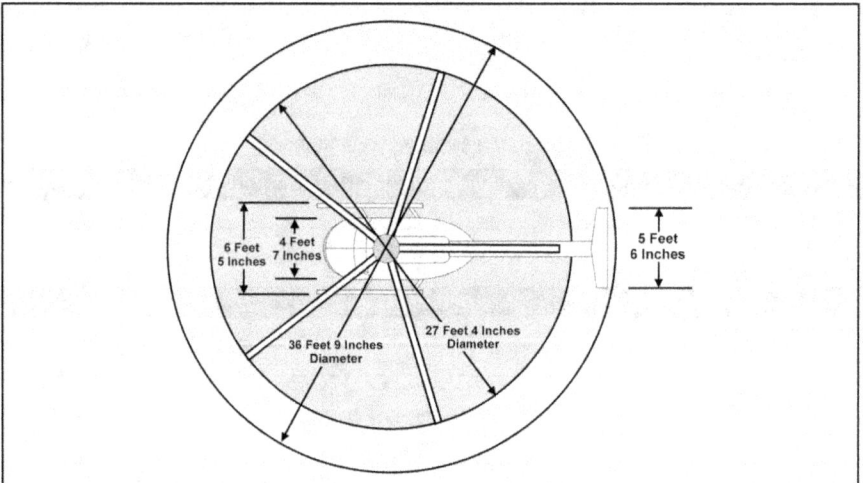

Figure B-3. MH-6M and AH-6M aircraft dimensions and turning radius

SAFETY

B-18. The MH-6M has no seat belts installed for passengers. Each passenger must provide his own means of securing himself. A short length of rope wrapped and knotted around the waist with a snap link attached to one end allows each passenger to secure himself to hard points on the aircraft.

AH-6M HELICOPTER

B-19. The mission of the AH-6M helicopter is to provide a rapidly deployable light attack helicopter to meet the need for surgical point and small-area target destruction or neutralization, with provisions for close air fire support for ground assault operations.

Aircraft Capabilities

DESCRIPTION

B-20. The AH-6M is a single engine light attack helicopter that has been modified to perform close air support of ground troops, target destruction raids, and armed escort of other aircraft. Its small size allows for rapid deployability in C-130, C-17, and C-5 transport aircraft. Aircraft modifications and aircrew training allow for extremely rapid upload and download times (Figure B-4).

Figure B-4. AH-6M helicopter

AIRCRAFT SURVIVABILITY EQUIPMENT

B-21. Other than the APR-39 Radar Warning Receiver System, the AH-6M currently is not equipped with aircraft survivability equipment. This passive omnidirectional warning set detects and identifies hostile search and acquisition and fire control radar. It provides audio and visual alerts to the flight crew. Additional survivability equipment includes fire extinguishers, first-aid kits, survival kits, and caving ladders. Caving ladders are installed as an emergency extraction device to recover other downed aircrews when operating overwater.

STANDARD MISSION EQUIPMENT

B-22. The AH-6M can have two Goliath tanks installed as an internal auxiliary fuel system. The tanks provide 62 additional gallons of fuel each. Each tank adds approximately 80 minutes of flight time. Additionally, each aircraft can have the following:

- *Forward-looking infrared AN/AAQ-16D, AESOP.* The Airborne Electro-Optical Special Operations Platform (AESOP) is a forward-looking infrared with a laser range designator. The AESOP allows the AH-6 to detect, acquire, identify, and engage targets at extended ranges with laser-guided munitions. The forward-looking infrared is a controllable, infrared surveillance system that provides a television-video-type infrared image of terrain features and ground or airborne objects of interest.
- *AN/ZSQ-3 forward-looking infrared.* This infrared imaging camera with image-intensified charge coupled device (I2CCD) camera coupled to forward-looking infrared provides a user selected "fused" images that can detect both hot spots and lights while minimizing infrared crossover limitations. It is gyro-stabilized with 360 degrees azimuth and elevation. It includes a laser pointer "sparkle" (steady and pulse), which is non-eye safe within 600 meters.

Appendix B

- *PACWIND receiver*. Real-time video feeds from intelligence, surveillance, and reconnaissance platforms can be displayed in the cockpit.

ARMAMENT

B-23. The AH-6M uses the "plank" system, which allows for multiple configurations. The plank features detachable outboard store stations, which are capable of folding and permit simplified aircraft transportability. Because of the flexibility of the plank system, multiple configurations of weapons systems are possible (Figures B-5 and B-6, page B-7). Provisions are available on the AH-6M plank system to mount and fire the following systems:

- *M134 7.62-millimeter (mm) Minigun*. The M134 minigun is a six-barrel, air-cooled, link-fed, electrically operated Gatling gun with a maximum effective range of 1,000 meters and tracer round burnout at 900 meters. The weapon has a rate of fire of 4,000 rounds per minute. The weapon fires a variety of 7.62-mm rounds: 7.62-mm ball, armor-piercing, and armor-piercing incendiary. Tracer mix is 4:1 for day operations. For night operations, 7.62-mm ball ammunition with a special low-light tracer is used. Night tracer mix is 9:1. Low-light tracer rounds prevent night vision goggles from shutting down during minigun engagements. A normal minigun load will have up to 500 rounds per gun. A heavy load magazine has from 2,250 rounds to 3,000 rounds per gun. Miniguns are normally mounted on inboard stores.
- *GAU-19 .50 Caliber Gatling Gun*. This weapon has a 1,000 rounds-per-minute rate of fire with a maximum effective range of 2,000 meters. The ammunition cans hold a maximum of 600 rounds of armor-piercing, armor-piercing incendiary, or Sabot-launched armor-piercing (SLAP) ammunition. The only low-light tracer ammunition available is an armor-piercing/armor-piercing incendiary mix. Up to 2 ammunition cans may be used for each gun; however, only 550 rounds may be loaded with a crossover, for a total of 1,100 rounds per gun. Tracer mix is 4:1 for both day and night. The AH-6 does not use .50 caliber ball ammunition because of ricochets. The GAU-19 can only be mounted on the outboard stores.
- *M260 Rocket Launcher*. This system fires 2.75-inch folding-fin aerial rockets with a variety of special-purpose warheads. There are 10-pound and 17-pound high-explosive warheads for light armor and bunker penetrations. The bursting radius for a 10-pound warhead is 10 meters, and 13 meters for the 17-pound warhead. The antipersonnel flechette warhead is filled with 1,179 80-grain flechettes. It has a minimum launch distance of 500 meters and an optimum launch distance of 1,000 to 3,000 meters. Other warheads include white and red phosphorus (used for smoke cover, marking targets, burning vegetation, and illumination). There are two types of illumination warheads; one type provides a bright light that can be seen with the naked eye (overt) and the other type provides a bright infrared (covert) light. Both are fired within 3,000 meters of the target area. After deployment, illumination warheads provide 120 seconds of overt light or 180 seconds of infrared light. The 2.75-inch rocket can be used as a point-target weapon at ranges from 100 to 1,000 meters and an area fire weapon at ranges up to 7,000 meters. The AH-6M can also fire inert rockets for training. M260 rocket pods may be mounted on inboard or outboard stores, but are normally mounted on the outboard stores. After landing, the AH-6M can be reloaded in 10 minutes.
- *M261 Modified 12-Tube Rocket Launcher*. This rocket launcher has modified 19-shot rocket pods which have the bottom two launcher rows removed. This system can fire both Mark 40 and Mark 66 rocket motors; 12 rocket pods may be mounted on the outboard stores only.
- *M261 19-Tube Rocket Launcher*. This rocket launcher has a 19-shot rocket pod. This system can fire both Mark 40 and Mark 66 rocket motors; 19 rocket pods may be mounted on the outboard stores only.
- *AGM-114 Hellfire*. The Hellfire is a 100-pound, semiactive laser-guided missile capable of defeating any known armor. Two M-272 Hellfire launchers hold four Hellfire missiles. The minimum engagement range is 0.5 kilometers, and the maximum range is 8 kilometers. The target can be laser-designated by any ground or air North Atlantic Treaty Organization standard laser designator.

Aircraft Capabilities

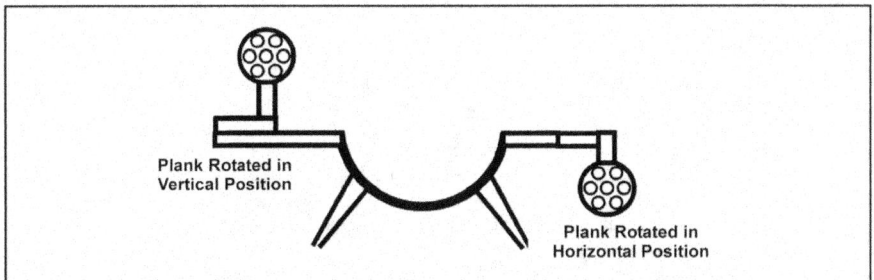

Figure B-5. AH-6M plank system for aircraft weapons configurations

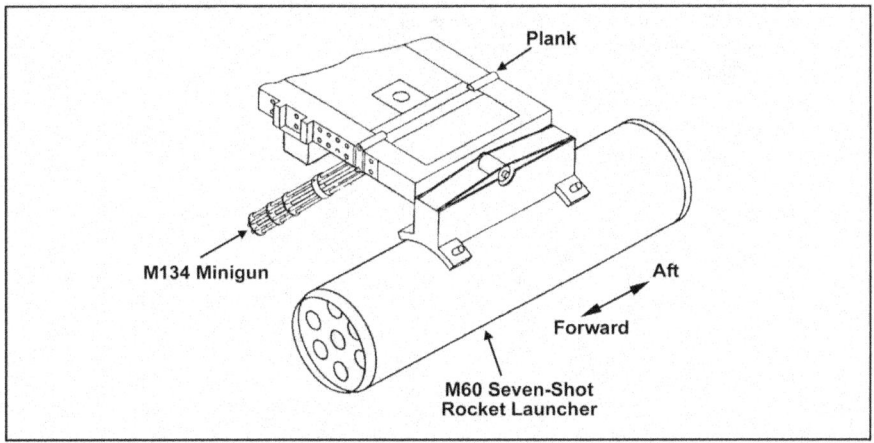

Figure B-6. AH-6M weapons variations

Weapons Employment

B-24. The pilot may fire weapons systems from either pilot station. He may fire rockets in singles (one at a time), pairs (two rockets, one from each rocket pod), or multiple rockets (depressing and holding down the firing button). The type of rocket to be fired is selectable from the cockpit, allowing the pilot to use the warhead that is applicable for the target.

B-25. During a call for fire, the forward controller can request a type of munition for a specific target (for example, a minigun only or flechettes), but the pilot has the final authority on munition types to be used during an engagement.

B-26. Minigun and 2.75-inch rocket targets include ground troops, buildings, small boats, aircraft, and thin-skinned vehicles (SLAP rounds can penetrate 3/4-inch homogeneous rolled steel). Hellfire missile targets include tanks and other hard-skinned vehicles, bunkers to some degree, larger boats, and buildings (the shaped warhead causes very localized damage). Normal engagement ranges are as follows:
- *Minigun*: 10 meters to 750 meters.
- *2.75-inch rockets*: 100 meters to 600 meters.
- *Hellfire missiles*: 500 meters to 8,000 meters.

Appendix B

TRANSPORTABILITY

B-27. The AH-6 can self-deploy or be transported in C-130, C-17, and C-5 aircraft. Self-deployment is unlimited with refuel support at ground or surface vessel locations every 270 nautical miles. Approximately 800 nautical miles per day is achievable, based on 8 hours of flight per day. Loading configurations on fixed-wing transport aircraft are as follows:
- *C-130*: Three AH-6s maximum administrative load and two AH-6s maximum for combat load.
- *C-17*: Nine AH-6s maximum administrative load and six AH-6s maximum for combat load.
- *C-5*: Twenty AH-6s maximum administrative load and thirteen AH-6s maximum for combat load.

B-28. In each case, tactical uploading and downloading of the aircraft can take place in an extremely short time. Off-load times vary, based upon numerous factors, such as ramp space, ramp condition, ramp type, off-load area, aircraft configuration, and mission configuration. General planning times for off-load from ramp down to takeoff (except C-5 deployment) are as follows:
- With the plank system, approximately 10 minutes.
- With the "T" tail removed, approximately 15 minutes.

PLANNING CONSIDERATIONS

B-29. The following paragraphs discuss considerations that must be taken into account when planning to use AH-6M aircraft in a mission.

Weather Minimums

B-30. A minimum 500-foot ceiling and 2-mile visibility capability must exist for day and night flying over flat or mountainous terrain or over water. The unit commander may reduce weather minimums on a mission-essential, case-by-case basis.

B-31. A visible horizon must exist in two of the four horizontal quadrants at all times. All AH-6M missions must take place under visual meteorological conditions rules.

Winds

B-32. The maximum wind allowed to start the aircraft is 40 knots, with a 20-knot gust spread.

Flight Altitudes

B-33. For training missions, the minimum en route altitude for routes not reconnoitered is 300 feet above ground level. The minimum overwater altitude is 50 feet. For operational missions, the minimum en route altitude is dependent upon mission, enemy, terrain and weather, troops and support available, time available, and civil considerations. The AH-6M can operate at altitudes ranging from 300 feet above ground level to 1500 feet above ground level in the terminal area. Higher altitudes are used at night for the purpose of noise reduction.

Landing Areas

B-34. The AH-6M is capable of landing on any structure that allows clearance for the rotor systems and meets stress requirements. Single-aircraft confined landing areas require a minimum size of 50 feet by 50 feet.

Shipboard Operations

B-35. The AH-6M can operate day and night from any ship having at least a one-spot helicopter-landing capability.

Aircraft Capabilities

Aircrew Composition

B-36. The aircrew of an AH-6M consists of two pilots—a pilot in command and a copilot. The pilot in command is responsible for the employment and actions of his aircraft. The copilot assists the pilot in command in accomplishing the mission. Both aircrew members have extensive training in navigation, gunnery, shipboard operations, overwater training, mountain flying, urban operations, and desert flying. The lead aircraft has a flight-lead-qualified pilot during all operations. The flight-lead pilot is responsible for mission accomplishment and is the primary mission planner.

Aircraft Capabilities

B-37. Table B-2 lists the capabilities of the AH-6M. The aircraft dimensions illustrated in Figures B-2 and B-3, page B-4, also pertain to the AH-6M.

Table B-2. AH-6M aircraft capabilities

Aircraft Weight	
Basic weight	2,196 pounds
Mission weight	3,100 pounds (fully fueled, dual-pilot)
Maximum gross weight	3,950 pounds
Aircraft Dimensions	
Length, blades folded	22 feet 6 inches
Length, blades unfolded	32 feet 1 inch
Width, blades folded	6 feet 5 inches
Width, blades unfolded	27 feet 4 inches
Height	8 feet 11 inches
Diameter of main rotor	27 feet 4 inches
Aircraft turning radius	36 feet 9 inches
Range and Endurance at 240 Pounds per Hour (With Optional Fuel Tank)		

Fuel Tank	Endurance (Hours + Minutes)	Fuel Range (Nautical Miles)
Main	1+17	116
Main plus one auxiliary	2+57	266

NOTE: Because of weight restrictions, the use of the optional fuel tank prevents the installation of a minigun and ammunition cans, and requires a reduced rocket load.

B-38. The cruise airspeed of the AH-6M is 90 knots indicated airspeed. The maximum airspeed is 108 knots indicated airspeed. All speeds are dependent on mission configuration and load.

Safety

B-39. Personnel must observe the following safety precautions:

- **Never** walk in front of armed aircraft.
- Wear protective headgear at all times when working around the turning rotor blades of the low rotor and tail rotor system of the AH-6M.
- Wear hearing and eye protection when working around operating aircraft.
- Be aware that the aircraft exhaust can start ground fires in extremely dry conditions with combustible material present (for example, dry grass or straw).
- Approach operating AH-6M aircraft as depicted in Figure B-7, page B-10.

Appendix B

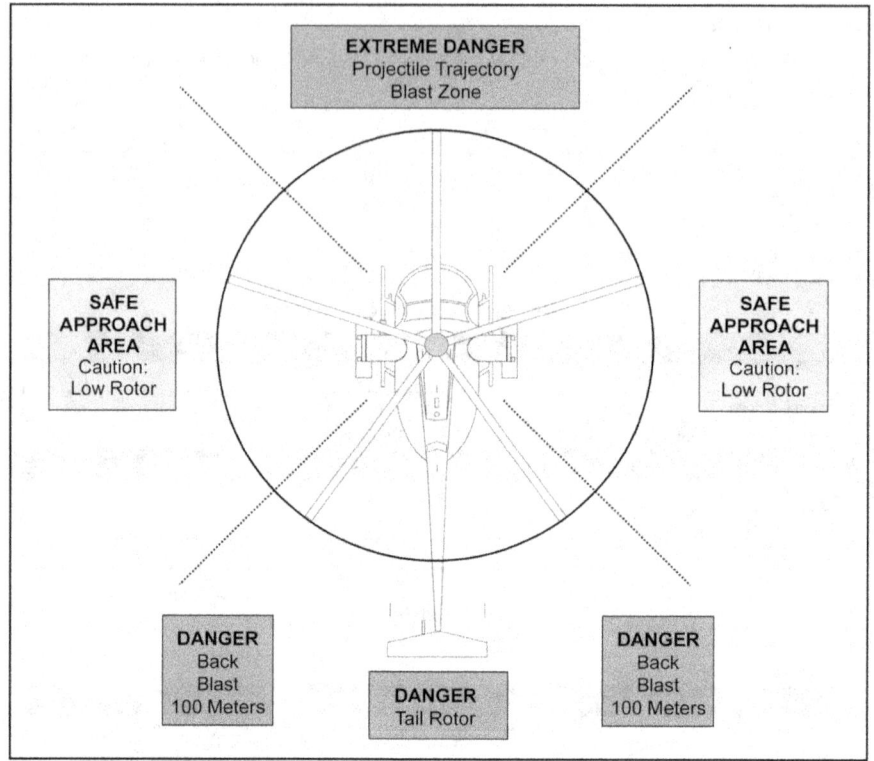

Figure B-7. AH-6M safety approach areas

MH-60L HELICOPTER

B-40. The primary mission of the MH-60L is to conduct overt and covert infiltration, exfiltration, and resupply of SOF across a wide range of environmental conditions and across the spectrum of conflict. Additionally, the MH-60L/DAP has the primary mission of armed escort and fire support. Secondary missions of the MH-60L include C2, external load, combat search and rescue, and medical evacuation operations. The MH-60L can operate from fixed-base facilities, remote sites, or oceangoing vessels.

DESCRIPTION

B-41. The MH-60L (Blackhawk) is a highly modified twin-engine utility helicopter (Figure B-8, page B-11). Its configuration may include a number of auxiliary fuel systems to allow for operational times of as much as 5.5 hours with a range of 640 nautical miles. The MH-60L has secure Selective Adaptive Communications Processor high frequency, FM, ultrahigh frequency, very high frequency, satellite, and Sabre communications. The FRIES allows for rapid insertion and extraction of personnel in areas blocked from air-land maneuvers. The aircraft has two M134 7.62-mm Gatling guns (miniguns), a ballistic armor subsystem, and aircraft survivability equipment to increase aircrew survivability in all threat environments. Dual GPSs, Omega, Doppler, and weather-detection systems allow pinpoint navigational and weather-avoidance capability. Mission-selective systems include a cargo hook for external load operations, a personnel locator system for combat search and rescue, and a four-place C2 console for airborne C2 operations. An armed version of the MH-60L, the defensive armed penetrator, is capable of mounting two M134 7.62-mm miniguns, two 30-mm chain guns, two 2.75 rocket pods, Hellfire missiles, or combinations of the systems for armed escort and fire support operations.

Figure B-8. MH-60L helicopter

AIRCRAFT SURVIVABILITY EQUIPMENT

B-42. The **AN/APR-39A (V) 1 Radar Warning Receiver System** identifies threat pulse radar in the C/D and H through J bands. It provides audio and visual alerts to the flight crew.

B-43. The **AN/APR-44 (V) 3 Radar Warning Receiver System** detects continuous wave surface-to-air missile threat radar emissions. It provides audio and visual warnings to the flight crew.

B-44. The **AN/ALQ-144 Infrared Countermeasures Set** provides false infrared signals to defeat the threat of infrared-sensing missiles.

B-45. The **M-130 Chaff Dispenser** dispenses decoy chaff as an effective countermeasure against radar-guided missiles.

B-46. The **AN/APX-100 (V) 1 IFF Transponder** provides automatic identification of the helicopter to suitably equipped ground and airborne interrogators.

B-47. The **emergency locator transmitter** transmits a distress signal on ultrahigh frequency and very high frequency guard frequencies. It goes on the avionics rack on the left side of the right pilot's seat. Impact activates the transmitter, or personnel may turn it on manually.

B-48. The **underwater acoustic beacon** radiates a pulsed acoustic signal of 37.5 kilohertz (kHz) detectable by hydrophone-equipped vessels. Water activates the beacon.

B-49. **Fire extinguishers** are hand-operated. One is mounted on the cabin wall forward of the right gunner's window and one is mounted on the right side of the left pilot's seat.

B-50. One **first-aid kit** is located on the back of the left pilot's seat, and one is mounted on the back of the right pilot's seat.

B-51. **Survival kits** are one or two environment-specific kits attached to the internal auxiliary fuel tanks.

Appendix B

STANDARD MISSION EQUIPMENT

B-52. The following paragraphs discuss standard equipment found on the MH-60L helicopter.

Armament

B-53. The standard armament is the M134 (7.62-mm minigun). The M134 is a six-barrel, air-cooled, electrically operated Gatling gun, with maximum effective fire of 1,000 meters. The gun fires A165 (7.62-mm balls), A257 (7.62-mm low-light balls), and SL66 (armor-piercing Sabots). One gun each is on the outside of the left and the right gunners' windows. The crew chiefs normally operate the guns, using open-steel, aim point, or aim-1 sights.

Ballistic Armor Subsystem

B-54. This item is a fabric-covered steel plating that provides increased ballistics protection in the cockpit and cabin.

Guardian Auxiliary Fuel Tanks

B-55. Two 172-gallon tanks, mounted in the cabin area at the aft bulkhead, provide range extension of approximately 2 hours (mains plus two auxiliary tanks, 4 hours total). Each tank occupies approximately 18 square feet of usable cabin floor space. Normal operational time without the guardian tanks is approximately 2 hours.

Fast-Rope Insertion and Extraction System Bar

B-56. Each side of the FRIES bar (Figure B-9) can support a maximum weight of 1,500 pounds.

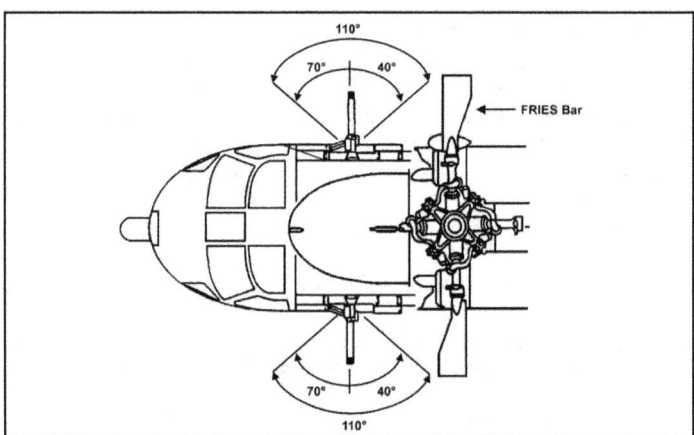

Figure B-9. MH-60L fast-rope insertion and extraction system bar

MISSION-SELECTIVE SYSTEMS

B-57. The following items are mountable on the MH-60L to support a primary mission or to enhance the capabilities of aircraft performing assault or DAP missions:
- *Cargo hook.* This item is mountable in the belly of the aircraft, below the main rotor. The hook can support external loads up to 9,000 pounds.

Aircraft Capabilities

- *External rescue hoist system.* This system is a hydraulic hoist capable of lifting 600 pounds. It contains 200 feet of usable cable. The crew chief or the hoist operator maneuvers the hoist using a handheld pendant.
- *Internal auxiliary fuel system.* The MH-60L has wiring provisions for four additional 150-gallon fuel cells, mountable in the cargo area. Each fuel cell provides approximately 50 minutes flight endurance. Ambient conditions and weight restrictions limit the maximum number of additional fuel cells. The use of all four internal auxiliary fuel system cells reduces usable cargo area space to near zero.
- *External extended-range fuel system.* This system consists of two 230-gallon, or two 230-gallon and two 450-gallon, or four 230-gallon jettisonable fuel tanks mountable on the external stores support system for long-range deployment of the aircraft. The use of the external extended-range fuel system restricts the employment of the M134 miniguns. Center-of-gravity or maximum-gross-weight restrictions and ambient conditions may limit the specific configuration of the extended-range fuel system.
- *C2 console.* This system provides four operator positions with access to the four AN/ARC-182 (V) multiband transceivers and forward-looking infrared display. Personnel may configure the MH-60L with an external stores support system to employ the family of loudspeakers—aircraft configuration (FOL-AC) with the supporting amplifier array frame on the cabin floor for MISO missions.

DEFENSIVE ARMED PENETRATOR SPECIAL CONFIGURATION

B-58. The mission of the armed MH-60L DAP (Figure B-10) is to conduct attack helicopter operations using area fire or precision-guided munitions and armed infiltration or exfiltration of small units. The DAP is a multimission aircraft capable of deploying on short notice and of conducting direct action missions. It is also capable of reconfiguring for troop assault operations. The DAP is capable of conducting all missions during day or night, or in adverse weather.

Figure B-10. MH-60L defensive armed penetrator

B-59. The DAP can provide armed escort for employment against threats to a vertical-lift formation. Using team tactics, the DAP is capable of providing close combat attack suppression or close air support for formations and teams on the ground. In the defensive armed role, the DAP is not a primary transport for troops or supplies because of high gross weights. The DAP conducting deep attacks has a combat radius of 225 nautical miles (takeoff, fly 225 nautical miles, no loiter, and return).

Appendix B

MH-60L Defensive Armed Penetrator Weapon System and Employment

B-60. The CMS-80 of the MH-60L DAP has integrated fire control systems. The integration gives the pilot a reduced cockpit workload and an increased weapons-selection capability through cockpit control driver and hands-on collective and stick weapons selection.

AN/AVQ-34 Monocular Heads-Up Display Set

B-61. The monocular heads-up display (MONOHUD) provides a lightweight, infinity focus, optical sight that allows the pilot to deliver rockets and gunfire effectively at targets. It gives the pilot cues for accurately launching missiles. It also provides aircraft flight symbology. The symbology is concise and provides all pertinent information in a manner that accommodates the pilot operating in daytime or with night vision goggles.

B-62. The standard armament configuration of the DAP is one rocket pod, one 30-mm cannon, and two miniguns. The configuration changes based on mission variables (Figure B-11).

Note: To avoid exceeding maximum gross weight limitations, reconfiguration of the ammunition or fuel mix may be necessary to achieve the desired insertion ranges for personnel when the MH-60L is in the DAP configuration.

External Fuel System Options

Left Outboard	Aircraft Window	Aircraft Window	Right Outboard
Rocket pod	Minigun	Minigun	Rocket pod
30-mm gun			30-mm gun
Hellfire missiles			Hellfire missiles
Fuel tank			Fuel tank
Stinger missiles			Stinger missiles

External Stores Support System Options

Left Outboard	Left Inboard	Aircraft Window	Aircraft Window	Right Inboard	Right Outboard
Rocket pod	Rocket pod	Minigun	Minigun	Rocket pod	Rocket pod
Hellfire missiles	30-mm gun			30-mm gun	Hellfire missiles
Fuel tank	Hellfire missiles			Hellfire missiles	Fuel tank
Stinger missiles	Fuel tank			Fuel tank	Stinger missiles

Figure B-11. Armament options for the MH-60L defensive armed penetrator

M134 7.62-mm Minigun

B-63. The M134 is a six-barrel, air-cooled, link-fed, electrically driven Gatling gun, with a 1,000-meter maximum effective range and a tracer burnout at 900 meters. The weapon has a rate of fire of 2,000 or 4,000 rounds per minute. The weapon is mountable in the fixed position on the left and right sides of the aircraft. The minigun fires a variety of 7.62-mm rounds. Nighttime operations use a 7.62-mm ball with a special low-light tracer, which prevents the shutting down of night vision goggles. The weapon also fires 7.62 SLAP ammunition for light-armor penetration. The DAP normally carries 3,000 rounds of 7.62-mm ammunition.

M261 19-Shot Rocket Launcher

B-64. The M261 fires a 2.75-inch folding-fin aerial rocket with a variety of special-purpose warheads. It has a 10-pound and a 17-pound high-explosive warhead for light-armor and bunker penetration. The bursting radius for the 10-pound warhead is 8 to 10 meters; for the 17-pound warhead, it is 12 to 15 meters. The antipersonnel flechette warhead contains 2,200 flechettes. Its minimum launch distance is 800 meters, and its optimum range is 1,100 meters. Another warhead is white phosphorous, used for smoke. The illumination warheads come in two types. One provides a bright light; the other, a bright infrared light. Firing of the illumination warheads is within 3,000 meters of the target area. After deploying, the warheads provide 120 seconds of overt light or 180 seconds of infrared light. The multipurpose submunition warhead contains nine submunitions that are effective against light armor and personnel. The multipurpose submunition round has a fuse that can be preset and that deploys the submunitions at the desired distance. The 2.75-inch rocket is useful as a point-target weapon at ranges from 100 to 750 meters and an area fire weapon at ranges up to 7,000 meters. The DAP can also fire chlorobenzaimalononitrile, high-explosive proximity, and inert rockets. The aircraft can carry an additional load of rockets internally, allowing the aircrew to reload the rocket pod. The aircrew can accomplish the reload within 15 minutes.

M230 30-mm Chain Gun

B-65. The M230 has its own magazine capable of carrying 1,100 rounds. The M230 has a cyclic rate of fire of 625, plus or minus 25 rounds per minute. The M230 is capable of firing the high-explosive dual-purpose, target practice, and target practice tracer. The high-explosive dual-purpose round is effective against light armor and personnel at ranges of 4,000 meters. With the use of the monocular heads-up display as a sighting system, the 30-mm cannon is a point-target weapon at a range of 1,500 meters or less. It is also an area fire weapon at ranges up to 4,000 meters.

AGM-114 Hellfire

B-66. The AGM-114 is a 100-pound semiactive laser-guided missile, capable of defeating any known armor. The M272 launchers are able to hold four Hellfire missiles each. The minimum engagement range is 0.8 kilometer; the maximum is 8 kilometers. Any ground or air North Atlantic Treaty Organization standard laser designator can designate the missile.

AN/AAQ-16D Airborne Electronic Special Operations Payload Forward-Looking Infrared

B-67. The AN/AAQ-16D is a forward-looking infrared with a laser range finder or designator. The AN/AAQ-16D allows the DAP to detect, acquire, identify, and engage targets at extended ranges with laser-guided munitions. The forward looking infrared is a controllable, infrared surveillance system that provides a television-video-type infrared image of terrain features and ground or airborne objects of interest. The forward-looking infrared is a passive system and detects long-wavelength radiant infrared energy emitted, naturally or artificially, by any object in daylight or darkness.

Air-to-Air Stinger

B-68. The DAP can fire the infrared seeking, fire-and-forget missile.

MH-60L DEFENSIVE ARMED PENETRATOR RECONFIGURATION

B-69. The MH-60L DAP has the capability to perform utility and armed missions. The time to reconfigure the aircraft from the armed to the utility or vice versa is minimal. The 7.62-mm miniguns remain with the aircraft regardless of the mission.

TRANSPORTABILITY

B-70. C-5A/B and C-17 aircraft can deploy the MH-60L, including the DAP configuration. The C-5A/B can carry a maximum of six MH-60Ls. The helicopters need a short time to prepare for on-load and again for rebuild upon arrival at the destination. The C-17 can carry three MH-60Ls.

PLANNING CONSIDERATIONS

B-71. Successful mission accomplishment is largely a function of adequate premission planning time. Mission notification should occur in time to have an adequate mission-planning session and briefing, followed by a period of rest before mission execution.

Appendix B

Weather Minimums

B-72. For training missions, forecast and actual weather requirements are a 500-foot ceiling and 2-mile visibility. For contingency missions, as directed by the commander, a 500-foot ceiling and 2-mile visibility work well for planning purposes. This type of forecast allows for en route cruise speed of the standard 120 knots and ample opportunity to adjust mission execution in the event of lower weather.

Winds

B-73. The MH-60L rotor has the capability to start and stop in actual winds no greater than 45 knots.

Flight Altitudes

B-74. For training missions, the minimum altitude is 300 feet above ground level for routes not reconnoitered and 150 feet above ground level for reconnoitered routes. For contingency missions, the minimum altitude is dependent upon mission, enemy, terrain and weather, troops and support available, time available, and civil considerations.

Landing Areas

B-75. The minimum landing area for the MH-60L is 100 feet by 100 feet.

Shipboard Operations

B-76. The MH-60L, including the DAP, can operate day and night from U.S. Navy ships with Level II, Class II helicopter-landing pads. For DAP, because of the high radio or radar electromagnetic interface signature onboard Navy vessels, only Mark 66 MOD-3 rocket motors are compatible with shipboard operations without waiver approval.

Aircrew Composition

B-77. Most training flights and all night vision goggle operations require four aircrew members—a pilot in command, a pilot, and two aircrew chiefs or gunners. One aircrew chief is at the right gunner's position. He scans for hazards, operates the hoist, conducts FRIES operations, operates the minigun, and conducts external load operations. The other aircrew chief is at the left gunner's position and scans for hazards, conducts FRIES operations, operates the minigun, and assists in external load operations.

Aircrew Qualifications

B-78. All aircrews are qualified to support flight operations for the missions stated in FM 3-05 and JP 3-05. Aircrew qualifications include multiship night vision goggle infiltration, exfiltration, and live-fire operations in urban, overwater, mountain, desert, jungle, and chemical, biological, radiological, and nuclear environments to landing zones, buildings, ships, and oil rigs. Aircrews are trained in night vision goggle long-range overland and overwater navigation, with an arrival standard of plus or minus 30 seconds.

Aircraft Capabilities

B-79. Table B-3, page B-17, lists the capabilities of the MH-60L.

Table B-3. MH-60L aircraft capabilities

Maximum Gross Weight (Ferry)	
Ferry configuration	23,500 pounds
Assault configuration	22,000 pounds
Aircraft Dimensions	
Length	64 feet 10 inches
Width	53 feet 8 inches
Height	16 feet 10 inches
Diameter of main rotor	53 feet 8 inches

Range and Endurance		
Fuel Tank	Endurance (Hours + Minutes)	Fuel Range (Nautical Miles)
Main	1+45	212
Main plus one auxiliary	3+02	364
Main plus two auxiliaries	4+10	496
Main plus three auxiliaries	5+00	600

Airspeed	
Cruise	120 knots indicated airspeed
Maximum	165 knots indicated airspeed

MH-60K HELICOPTER

B-80. The primary mission of the MH-60K (Blackhawk) is to conduct overt or covert infiltration, exfiltration, and resupply of SOF over a wide range of environmental conditions. The MH-60K is capable of operating from fixed-base facilities, remote sites, or oceangoing vessels.

DESCRIPTION

B-81. The MH-60K is a highly modified twin-engine utility helicopter (Figure B-12, page B-18). It is equipped with weather-avoidance and search radar, an aerial refueling probe for in-flight refueling, and a forward-looking infrared sensor that displays inside the cockpit, enhancing mission capability and safety. The aircraft can be configured with a number of auxiliary fuel systems to allow for unrefueled operational times of as much as 4.0 hours with a range of 440 nautical miles. The MH-60K is equipped with secure high frequency, single-channel ground and airborne radio system (SINCGARS), FM, ultrahigh frequency, very high frequency, and two multiband radios, each capable of satellite communications, or line-of-sight communications. The FRIES allows for rapid insertion and extraction of personnel in areas that preclude aircraft from landing. The aircraft has two M134 7.62-mm Gatling guns, a ballistic armor subsystem, and an aircraft survivability equipment suite to increase aircrew survivability in all threat environments. The GPS, inertial navigation unit, and attitude and heading reference system allow pinpoint navigation. The multimode radar system is a mission aid and allows for near-zero-visibility penetration under some situations. Mission-selective systems include the cargo hook for external load operations and the personnel locator system for combat search and rescue. The MH-60K helicopter is instrument-capable with an automatic direction finder, very high frequency omnidirectional range, distance-measuring equipment, instrument landing system, and tactical air navigation. Mission-computer-generated approaches can be used when normal approaches are unavailable.

Appendix B

Figure B-12. MH-60K helicopter

AIRCRAFT SURVIVABILITY EQUIPMENT

B-82. The **AN/APR-39A (V) 1 Radar Warning Receiver** identifies threat pulse radar in the C or D and the H through J bands. It provides audio and video alerts to the flight crew.

B-83. The **AN/APR-44 (V) 3 Radar Warning Receiver System** detects continuous wave surface-to-air missile threat radar emissions. It provides audio and visual warnings to the flight crew.

B-84. The **AN/AVR-2 Laser Warning Receiver** detects laser emissions directed toward the helicopter.

B-85. The **AN/AAR-47 Missile Warning Receiver** detects plume emissions from a missile's exhaust.

B-86. The **AN/ALQ-162 (V) 2 Continuous Wave Radar Jammer** detects and jams continuous wave radar emitters.

B-87. The **AN/ALQ-136 (V) 2 Pulse Radar Jammer** detects and jams pulse radar emitters.

B-88. The **AN/ALQ-144 Infrared Countermeasures Set** provides false infrared signals to defeat infrared-sensing missile threats.

B-89. The **AN/AAR-57 Common Missile Warning System** allows for automatic or manual dispensing of decoy chaff and flare as an effective countermeasure against radar-guided and infrared missile threats.

B-90. The **AN/APX-118 IFF Transponder With Mode S** provides automatic identification of the helicopter to suitably equipped ground and airborne interrogators.

B-91. The **emergency locator transmitter** is mounted on the left side of the right pilot's seat avionics rack. It transmits a distress signal on ultrahigh frequency and very high frequency guard frequencies. The transmitter is impact-activated but may be turned to the ON state manually.

B-92. The **underwater acoustic beacon** is activated by contact with water. It radiates a pulsed acoustic signal of 37.5 kilohertz detectable by hydrophone-equipped vessels.

B-93. **Fire extinguishers** are hand-operated—one is mounted on the cabin wall forward of the right gunner's window and one is mounted on the right side of the left pilot's seat.

B-94. **Two first-aid kits** are located on the back of the left pilot's seat and one is mounted on the back of the right pilot's seat.

B-95. **Survival kits** are environment-specific kits attached to the internal auxiliary fuel tanks.

Aircraft Capabilities

STANDARD MISSION EQUIPMENT

B-96. The following are systems and equipment always on board the aircraft during tactical missions. This list does not include avionics, aircraft survivability equipment, and sensors, as they are considered part of the basic aircraft.

Armament

B-97. The standard armament for the MH-60K is the M134 (7.62-mm minigun), six-barrel, air-cooled, and electrically operated Gatling gun (Figure B-13). The maximum effective range is 1,000 meters. The M134 fires A165 (7.62 ball), A257 (7.62 low-light ball), and SL66 (armor-piercing Sabot) ammunition. One gun is mounted outside both the left and right gunner's windows. Aircrew chiefs normally operate the weapon system. Weapon sighting is by open steel sights.

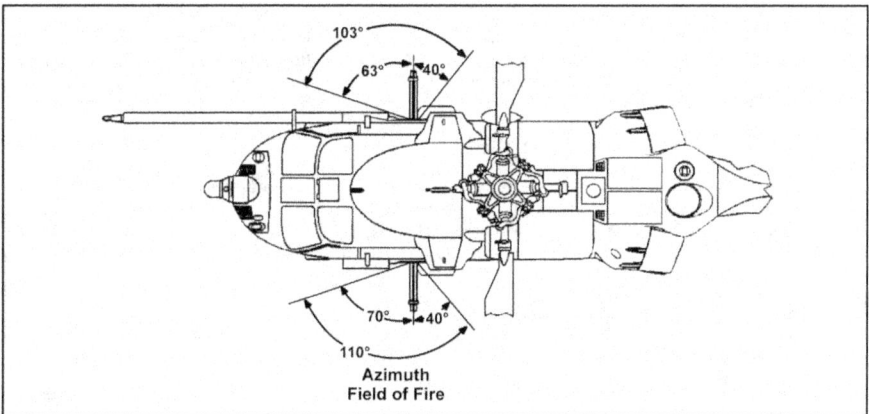

Figure B-13. MH-60K M134 minigun window-mounted field of fire

Ballistic Armor Subsystem

B-98. The system consists of fabric-covered steel plating, which provides increased ballistic protection in the cockpit and cabin areas.

Robinson-Guardian and Internal Auxiliary Fuel Tank Systems

B-99. The option of installing one or two auxiliary fuel tanks is available. The single 200-gallon tank can extend range/station time by approximately 1 hour (2.5 hours total). The two 172-gallon fuel tanks system provides a range extension of approximately 2 hours (3.5 hours total). The tanks are mounted in the cabin area at the aft bulkhead. The single 200-gallon tank occupies approximately 11 square feet, and the two 172-gallon tanks system occupies approximately 18 square feet of usable cabin floor space. If tanks are empty, approximately 1 hour is needed to remove each auxiliary tank system.

Fast-Rope Insertion and Extraction System Bar

B-100. Each side of the FRIES bar can support a maximum weight of 1,500 pounds.

Appendix B

Mission-Selective Systems

B-101. The following systems are mountable on the MH-60K to support a primary mission or to enhance the capabilities of aircraft performing assault missions:

- *Cargo hook.* This system is mountable in the belly of the aircraft below the main rotor. It can support external loads up to 8,000 pounds.
- *External rescue hoist.* This hydraulic hoist is capable of lifting 600 pounds with 200 feet of usable cable. The aircrew chief or hoist operator uses a handheld pendant to control the system.
- *Aerial refueling system.* This system is an aerial refueling probe that allows extended range and endurance by refueling from HC/MC/KC-130 tanker aircraft.
- *Aerial loudspeaker system.* The MH-60K with external stores support system can be configured to employ the 2,700-watt family of loudspeakers—aircraft configuration with the supporting amplifier array frame on the cabin floor for MISO aerial loudspeaker missions.
- *Micro-forward area refueling equipment system.* The forward area refueling equipment consists of fueling pumps, hoses, nozzles, and additional refueling equipment to set up a two-point refueling site. Gallons of fuel dispensed are dependent upon the range of operation required of the tanker aircraft.

Transportability

B-102. The MH-60K may be deployed by C-5A, C-5B, and C-17 aircraft. A maximum of six MH-60Ks can be loaded on a C-5A or a C-5B. A short time is needed to prepare the helicopters for on-load and again for rebuilding on arrival at the destination. Three MH-60Ks can be loaded onto a C-17. Ammunition for the weapon systems is palletized and loaded on the same aircraft for distribution at the destination.

Planning Considerations

B-103. Successful mission accomplishment is largely a function of adequate premission planning. Mission notification should occur in time for an adequate mission-planning session and briefing, followed by a period of rest before execution.

Weather Minimums

B-104. Training missions require forecast and actual weather parameters of a 500-foot ceiling and 2-mile visibility. For contingency missions, as directed by the commander, a 500-foot ceiling and 2-mile visibility work well for planning purposes. This type of forecast allows for en route cruise speed of the standard 110 knots and ample opportunity to adjust mission execution depending upon weather conditions. When en route weather conditions are less than a 500-foot ceiling with 2-mile visibility, tactical instrument meteorological conditions and/or terrain-following radar may be used within certain constraints and planning considerations.

Flight Altitudes

B-105. For training missions, the minimum altitude is 300 feet above ground level for routes not reconnoitered and 150 feet above ground level for reconnoitered routes. Contingency missions are dependent upon mission, enemy, terrain and weather, troops and support available, time available, and civil considerations.

Landing Areas

B-106. The minimum landing area for the MH-60K is 100 feet by 100 feet. For shipboard operations, the MH-60K can operate day and night from Navy ships that have Level II, Class II helicopter-landing pads.

Aircraft Capabilities

Aircrew Composition

B-107. Most training flights and all night vision goggle operations require four aircrew members. These members include a pilot in command, a pilot, and two aircrew chiefs or gunners. One aircrew chief—stationed at the right gunner's position—scans for hazards, operates the hoist, conducts FRIES operations, operates the minigun, and conducts external load operations. The other aircrew chief—stationed at the left gunner's position—conducts FRIES operations, operates the minigun, and assists in external load operations.

Aircrew Qualifications

B-108. Aircrews can perform all mission tasks in all environments. They can perform night vision goggle long-range overland and overwater navigation, with an arrival standard of plus or minus 30 seconds.

Aircraft Capabilities

B-109. Table B-4 lists the capabilities of the MH-60K aircraft. Figures B-14 through B-16, pages B-22 through B-24, illustrate specific dimensions and capabilities of the aircraft.

Table B-4. MH-60K aircraft capabilities

Aircraft Weight		
Basic weight	15,600 pounds	
Maximum gross weight	24,500 pounds	
Aircraft Dimensions		
Length, blades folded	60 feet 7 inches	
Length, blades unfolded	64 feet 10 inches	
Width, blades folded	9 feet 9 inches	
Width, blades unfolded	53 feet 8 inches	
Height	16 feet 10 inches	
Diameter of main rotor	53 feet 8 inches	
Aircraft turning radius	41 feet 8 inches	
Range and Endurance		
Fuel Tank	Endurance (Hours + Minutes)	Fuel Range (Nautical Miles)
Main	1+30	165
Main plus one auxiliary	2+30	275
Main plus two auxiliaries	3+30	385
Airspeed		
Cruise	115 knots indicated airspeed	
Maximum	145 knots indicated airspeed	

Appendix B

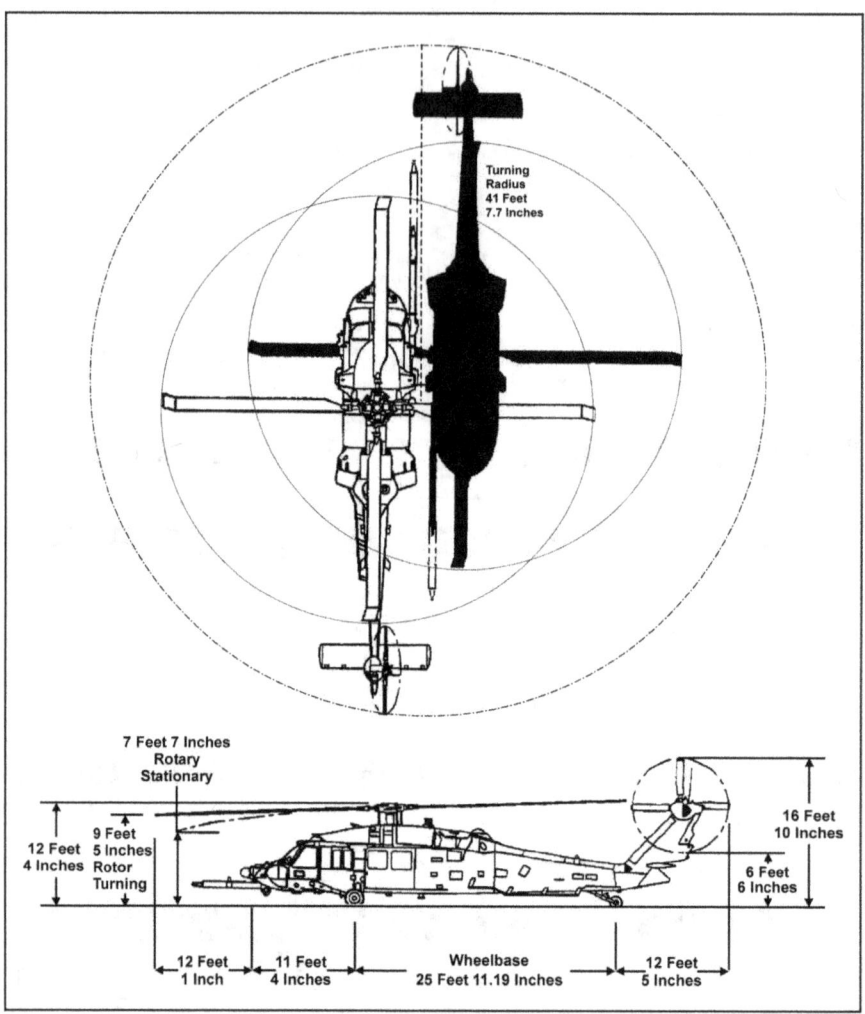

Figure B-14. MH-60K dimensions and turning radius

Aircraft Capabilities

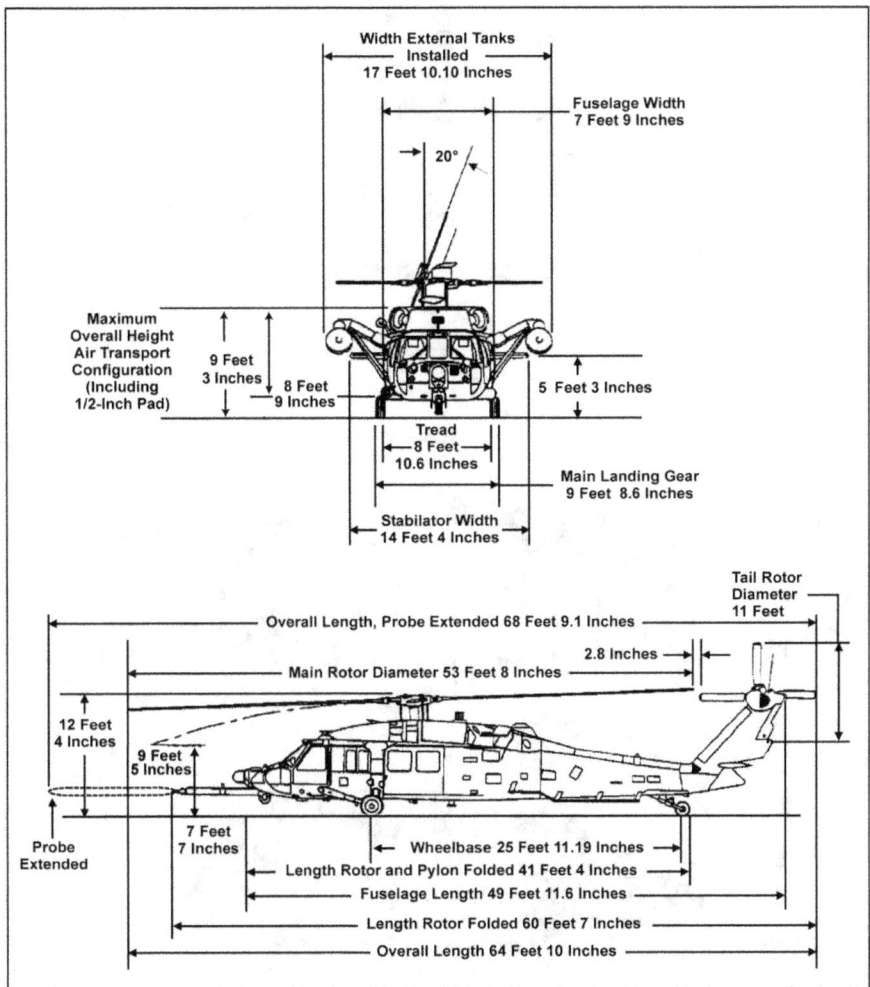

Figure B-15. MH-60K dimensions for intertheater airlift preparation

Appendix B

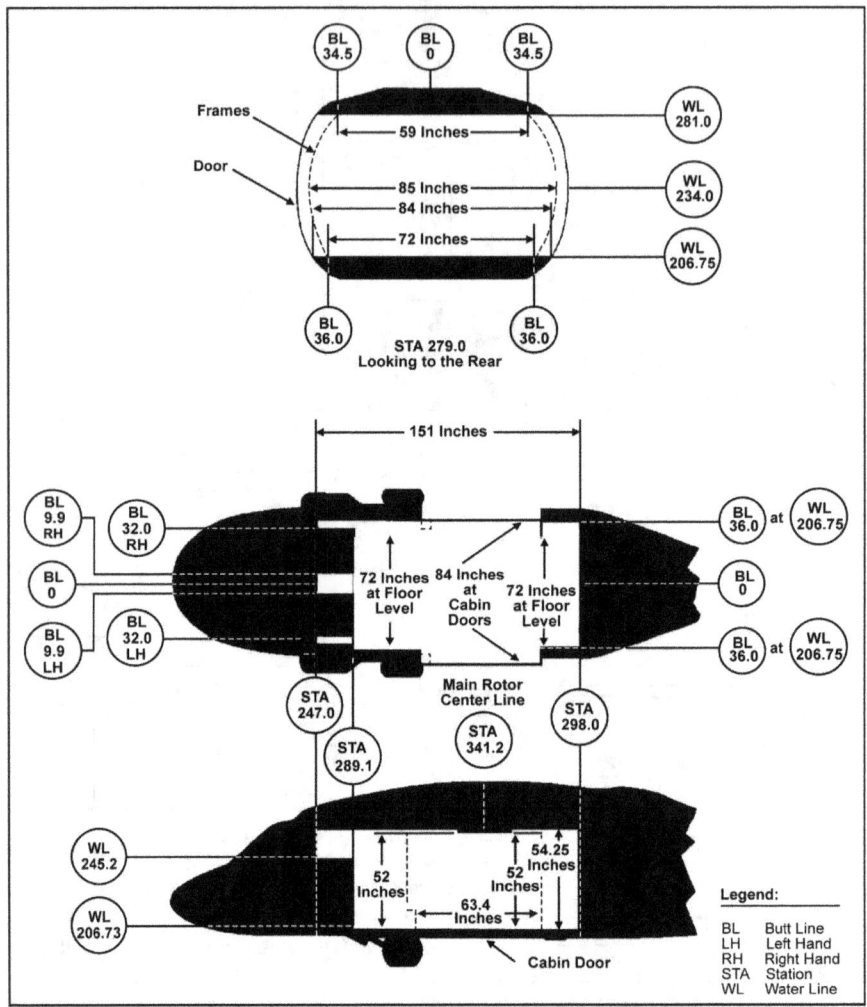

Figure B-16. MH-60K aircraft capabilities

MH-47G HELICOPTER

B-110. The primary mission of the MH-47G is to conduct overt and covert infiltration, exfiltration, air assault, resupply, and external-sling operations under a wide range of environmental conditions. The aircraft can perform a variety of other missions, including shipboard, platform, urban, water, forward arming and refueling point, mass-casualty, and combat search and rescue operations.

Aircraft Capabilities

Description

B-111. The MH-47G (Chinook) is a twin-engine, tandem-rotor, heavy assault helicopter specifically designed and built for the SOA mission (Figure B-17). It has a totally integrated avionics subsystem that combines the following:
- Redundant common avionics architecture system with dual mission processors.
- Data concentrator units.
- Multifunction displays and display generators to improve combat survivability and mission reliability.
- Aerial refueling probe for in-flight refueling.
- External rescue hoist.
- Two L714 turbine engines with full-authority digital electronic control.
- Two integral aircraft fuel tanks providing 2,068 gallons of fuel.
- Stormscope for thunderstorm avoidance.
- Multi-Mode Radar for terrain-following flight.

Figure B-17. MH-47G helicopter

Aircraft Survivability Equipment

B-112. The **AN/APR 39A Radar Warning Receiver** identifies hostile pulse fire control radar and provides audio and video alerts to the flight crew when the system detects threat radar emissions.

B-113. The **AN/ALE-47 (V) Countermeasures Dispensing System** consists of five components used to provide preemptive and terminal threat protection. The pilots control the system by using the cockpit control unit mounted in the center console. The AN/ALE-47 replaces the M-130 system and enhances aircraft survivability by—
- Integrating with avionics and electronic warfare systems.
- Providing threat-adaptive programmable dispensing routines.
- Providing data links for advanced expendables.
- Using available threat sensors.

Appendix B

B-114. The **AN/AAR-57 Common Missile Warning System** is a passive electronic warfare system that detects in-band infrared and ultraviolet radiation emanating from a missile plume.

B-115. The **AN/APR 44 Radar Warning Receiver** identifies hostile airborne interceptor and surface-to-air missile continuous-wave fire control radar, and provides audio and visual alerts to the flight crew when the system detects threat radar emissions.

B-116. The **Improved Infrared Countermeasures Dispenser System** is a chaff and flare dispenser system designed to deceive radar guidance and infrared missiles by using chaff or flares as required.

STANDARD MISSION EQUIPMENT

B-117. The following paragraphs discuss standard equipment found on the MH-47G helicopter.

Armament

B-118. The MH-47G has four weapon stations—left forward window, right cabin door, and at the aft left and right windows. The forward stations mount 7.62-mm miniguns, and the aft stations each mount an M240 7.62-mm machine gun. An aircrew member at each station manually operates the weapons. The primary use of the weapon is self-defense and enemy suppression. The minigun is normally used for soft targets and troop suppression, which requires a high rate of fire. The minigun is air-cooled and link-fed. It has a maximum effective range of 1,500 meters, with a tracer burnout at 900 meters. The weapon has an adjustable rate of fire of 2,000 or 4,000 rounds per minute. The aircrew members currently fire ball or SLAP ammunition with a mix of four ball rounds to one tracer round (4:1) or a 9:1 mix to prevent night vision device shutdown on low-illumination nights. The ammunition complement without reloading is 8,000 rounds per minigun and 1,500 rounds per M240.

Fast-Rope Insertion and Extraction System Bar

B-119. The FRIES is used for insertion and extraction of personnel. Applied loads for the FRIES are as follows:

- Applied loads at the rear ramp for insertions cannot exceed nine persons per rope at the same time.
- Applied loads at the rear ramp for extractions cannot exceed six persons per rope at the same time.

Map Display Generator

B-120. The map display generator displays aeronautical charts, photos, or digitized maps in the mission planning and 3D modes of operation based on digital terrain elevation data and digital feature analysis data.

MISSION-SELECTIVE SYSTEMS

B-121. The external cargo hook systems may be used separately or in conjunction with one another (Table B-5). All loads should be planned as a tandem-rigged load to facilitate greater load stability and to ensure faster airspeeds during flight.

Table B-5. MH-47G external cargo hooks

Type	Capacity
Forward hook	20,000 pounds
Center hook	28,000 pounds
Aft hook	20,000 pounds
Tandem hook	25,000 pounds
NOTE: These are maximum hook-rated loads and may not accurately reflect the true capability of the aircraft because of external conditions, such as pressure, altitude, and temperature.	

B-122. The **external rescue hoist system** is configured for use at the right forward cabin door. It has a 600-pound capacity and approximately 200 feet of usable cable. Fast-rope operations can still be conducted out the front cabin door with the hoist installed.

B-123. The **internal rescue hoist** is for use at the center cargo hook and rescue hatch. It has a 600-pound capacity, with approximately 150 feet of usable cable.

B-124. The **AN/AAQ-16 forward-looking infrared** is a controllable, infrared surveillance system that provides a television-video-type infrared image of terrain features and ground or airborne objects of interest. The forward-looking infrared is a passive system and detects long-wavelength radiant infrared energy emitted, naturally or artificially, by any object in daylight or darkness.

B-125. The **cargo compartment expanded range fuel system** consists of one and up to three ballistic-tolerant, self-sealing tanks. Each tank has the capacity of holding 800 gallons of fuel but normally is filled to 780 gallons. Filling may occur during ground or aerial refueling operations.

B-126. The **forward area refueling equipment** consists of fueling pumps, hoses, nozzles, and additional refueling equipment to set up a two-point refueling site. Gallons of fuel dispensed are dependent upon the range of operation required of the tanker aircraft.

TRANSPORTABILITY

B-127. The SOAR has modified and validated procedures to load two MH-47Gs on a C-5 and one MH-47G on a C-17, with all support equipment. The time required to disassemble and assemble the MH-47G is greater than the time required of other SO helicopters.

PLANNING CONSIDERATIONS

B-128. Successful mission accomplishment is largely a function of adequate premission planning time. Mission notification should occur in time to have an adequate mission-planning session and briefing, followed by a period of rest before mission execution.

Weather Minimums

B-129. The weather minimum for the MH-47G is a 500-foot ceiling, with a visibility of 2 miles.

Winds

B-130. The MH-47G has no specified minimum winds for training, operational, and support missions; however, the maximum wind for starting and stopping the rotor system is 45 knots.

Flight Altitudes

B-131. For training missions, the minimum en route altitude for reconnoitered routes is 150 feet above ground level or above the highest obstruction. For routes not reconnoitered, the minimum en route altitude is 300 feet above ground level or above the highest obstruction. For operational missions, the minimum en route altitude is dictated by threat systems.

Landing Areas

B-132. The minimum landing area for the MH-47G is 150 feet by 100 feet.

Shipboard Operations

B-133. The MH-47G can operate day and night from Navy ships that have Level II, Class III helicopter-landing pads.

Appendix B

Aircrew Composition

B-134. Most training, exercises, and operational or contingency missions require five aircrew members, including a pilot, a copilot, a flight engineer, and two aircrew chiefs. The flight engineer, usually positioned at the ramp station, scans for other aircraft, targets, and obstacles. He also operates the hoist (when required), assists in FRIES operations, operates the machine gun, and conducts sling-load operations. The aircrew chiefs, positioned at the left and right forward gunners' stations, scan for other aircraft, targets, and obstacles. They also operate the miniguns and assist in sling-load and FRIES operations.

Aircrew Qualifications

B-135. MH-47G aircrews can support flight operations for the missions stated in FM 3-05 and JP 3-05. Aircrew qualifications include night vision goggle infiltration and exfiltration operations to urban, overwater (ship, oil rigs), mountainous, desert, and jungle objectives arriving at the target at a prearranged time plus or minus 30 seconds. Aircrews are trained in formation live-fire, long-range night vision device operations over land and water. MH-47G aircrews can also perform aerial-refueling operations.

Aircraft Capabilities

B-136. Table B-6, pages B-28 and B-29, lists the capabilities of the MH-47G aircraft.

Table B-6. MH-47G aircraft capabilities

Aircraft Weight	
Maximum gross weight	54,000 pounds
Empty gross weight	29,000 pounds
Aircraft Dimensions	
Length of fuselage	52 feet 1 inch
Length of aircraft with probe	68 feet 5 inches
Length with blades (turning)	99 feet
Width of fuselage	15 feet 8 inches
Width with blades (turning)	60 feet
Height	18 feet 7 inches
Aircraft turning radius (ground handling)	122 feet
Cargo Area (Unobstructed)	
Height	78 inches
Width	90 inches
Depth	366 inches
Troop Capacity	
With seats	33 troops
Floor loading	65 troops
Litters	24 troops
Airspeed	
Normal cruise	120 knots indicated airspeed
Maximum dash	170 knots indicated airspeed
Maximum altitude	20,000 feet

Aircraft Capabilities

Table B-6. MH-47G aircraft capabilities (continued)

Fuel Flow	
Normal fuel consumption	2,750 pounds per hour
Maximum fuel consumption	3,300 pounds per hour
Range and Endurance		
Fuel Tank	Endurance (Hours + Minutes)	Fuel Range (Nautical Miles)
Integral	4+30	540
Aircrew endurance and aerial refueling support may limit the range.		
NOTE: Actual figures are dependent upon temperature, aircraft gross weight, and density altitude.		

Typical Mission Composition

B-137. A variety of mission scenarios may employ the MH-47G. A typical mission profile for a low-to-medium threat infiltration and exfiltration sortie could take the following form:

- Departing by night, multiship, visual meteorological conditions from a forward operating location to a target 520 nautical miles (range dependent upon fuel configuration and availability of prestaged forward arming and refueling point locations). The MH-47G aircraft is capable of aerial refueling. Lack of availability of KC-, HC-, or MC-130 support would limit the range.
- Navigating to an initial point using the best option of three navigational modes. GPS is the primary mode when the required number of satellites is available.
- Using guidance cues, following navigation steering to a landing site and accomplishing the approach to landing to a remote site of not less than 150 feet by 100 feet.
- On-loading an exfiltration party of up to 65 passengers from conditions not worse than 50 degrees Celsius (122 degrees Fahrenheit) at 500-foot pressure altitude.
- Reversing the route and returning to a recovery base or location in friendly territory.
- Using low en route altitudes (down to 50 feet) given favorable conditions of ambient light, visibility, and the use of infrared searchlight with night vision devices during adverse weather conditions.

Capabilities Matrix

B-138. The SOAR aircraft capabilities matrix (Table B-7, pages B-29 through B-31) is a ready reference for mission planners. Its purpose is to reduce mission-planning time. The matrix provides instant information, without time-consuming research on the part of mission planners. The current aircraft capabilities matrix encompasses all aircraft systems. SOAR units must update the matrix periodically as technology changes occur.

Table B-7. SOAR aircraft capabilities matrix

	MH-6M	AH-6M	MH-60L	MH-60L (DAP)	MH-60K	MH-47G
Aircraft Capabilities						
Cruise Speed (Knots)	80	90	120	120	115	120
Flight Time (Standard Tanks)	1+00	1+17	1+45	1+40	1+30	4+30
Range (Nautical Miles) (Standard Tanks)	80	116	212	200	165	540
Air-Refuelable	No	No	Yes	Yes	Yes	Yes
Passengers (not including crew)	6	0	12	0	14	33

Appendix B

Table B-7. SOAR aircraft capabilities matrix (continued)

	MH-6M	AH-6M	MH-60L	MH-60L (DAP)	MH-60K	MH-47G
Maximum Passengers (No Seats for Rucksacks)	6	0	17	0	14	33
Landing Area (Feet)	50x50	50x50	100x100	100x100	100x100	150x100
Communications						
Ultrahigh Frequency	Yes	Yes	Yes	Yes	Yes	Yes
High Frequency	No	No	Yes	Yes	Yes	Yes
Very High Frequency (FM/AM)	Yes	Yes	Yes	Yes	Yes	Yes
Satellite Communications	Yes	Yes	Yes	Yes	Yes	Yes
SINCGARS	Yes	Yes	Yes	Yes	Yes	Yes
Sabre	Yes	Yes	Yes	Yes	Yes	Yes
Airborne Target Handover System	No	No	Yes	Yes	No	No
Navigation						
Instrument Meteorological Conditions Certified	No	No	Yes	Yes	Yes	Yes
GPS	Yes	Yes	Yes	Yes	Yes	Yes
Automatic Direction Finder	No	No	Yes	Yes	Yes	Yes
Very High Frequency Omnidirectional Range/ Distance-Measuring Equipment	Yes	Yes	Yes	Yes	Yes	Yes
Instrument Landing System	No	Yes	Yes	Yes	Yes	Yes
Doppler	No	No	Yes	Yes	Yes	Yes
Inertial Navigation System	No	No	No	No	Yes	Yes
Attitude Heading Reference System	No	No	No	No	Yes	No
Tactical Air Navigation	Yes	Yes	Yes	Yes	Yes	Yes
Air Data Computer	No	No	No	No	No	Yes
Personnel Locator System	No	No	Yes	Yes	Yes	Yes
Mission Computer Unit	No	No	No	No	Yes	Yes
Long-Range Navigation (LORAN)	Yes	Yes	No	No	No	No
Armament						
M134 7.62-mm Minigun (Maximum)	No	Yes	Yes (2)	Yes (2)	Yes (2)	Yes (2)
M230 30-mm Chain Gun	No	No	No	Yes	No	No
M240 762-mm Machine Gun	No	No	No	No	No	Yes (2)
M260 7-Shot Rocket	No	Yes	No	Yes	No	No
M261 19-Shot Rocket	No	Yes	No	Yes	No	No
AGM-114 Hellfire (Maximum)	No	Yes (4)	No	Yes (16)	No	No

Aircraft Capabilities

Table B-7. SOAR aircraft capabilities matrix (continued)

	MH-6M	AH-6M	MH-60L	MH-60L (DAP)	MH-60K	MH-47G
M2 Caliber .50 Machine Gun (Maximum)	No	Yes (2)	No	No	No	No
Standard of Special Equipment						
Ballistic Armor Subsystem	Yes	Yes	Yes	Yes	Yes	Yes
Guardian Auxiliary Fuel Tank	No	No	2+00	2+00	2+00	No
FRIES	Yes	No	Yes	Yes	Yes	Yes
Forward-Looking Infrared	Yes	Yes	Yes	Yes	Yes	Yes
External Cargo Hook	No	No	9,000 Pounds	No	8,000 Pounds	28,000 Pounds
Rescue Hoist	No	No	600 Pounds	No	600 Pounds	600 Pounds
Auxiliary Fuel System (Time/Nautical Miles)	3+00/240	2+57/266	5+00/600	4+45/570	3+30/385	8+20/1000
C2 Console	No	No	Yes	No	No	Yes

This page intentionally left blank.

Glossary

The Glossary lists acronyms and terms with Army, multi-Service, or joint definitions, and other selected terms. Terms for which FM 3-76 is the proponent manual (the authority) are marked with an asterisk (*). The proponent manual for other terms is listed in parentheses after the definition.

SECTION I – ACRONYMS AND ABBREVIATIONS

A	airborne
ALE	Army special operations forces liaison element
AM	amplitude modulation
ARSOAC	Army Special Operations Aviation Command
ARSOF	Army special operations forces
C2	command and control
CONUS	continental United States
DAP	defensive armed penetrator
FID	foreign internal defense
FM	field manual; frequency modulation
FRIES	fast-rope insertion and extraction system
G-2	Deputy Chief of Staff for Intelligence
G-3	Deputy Chief of Staff for Operations and Plans
G-8	Deputy Chief of Staff for Resource Management
GPS	Global Positioning System
H-hour	specific time an operation or exercise begins
HQ	headquarters
IFF	identification, friend or foe
J-4	logistics directorate of a joint staff
JFSOC	joint force special operations component
JFSOCC	joint force special operations component commander
JP	joint publication
JSOA	joint special operations area
JSOAC	joint special operations air component
JSOACC	joint special operations air component commander
JSOTF	joint special operations task force
MDMP	military decisionmaking process
MISO	Military Information Support operations
mm	millimeter(s)
MOS	military occupational specialty
S-2	intelligence staff officer
S-3	operations staff officer
S-4	logistics staff officer
S-6	command, control, communications, and computer operations officer

Glossary

SB(SO)(A)	Sustainment Brigade (Special Operations) (Airborne)
SEAL	sea-air-land team
SF	Special Forces
SFODA	Special Forces operational detachment A
SINCGARS	single-channel ground and airborne radio system
SLAP	Sabot-launched armor-piercing
SO	special operations
SOA	special operations aviation
SOAR	Special Operations Aviation Regiment
SOCCE	special operations command and control element
SOCM	special operations combat medic
SOF	special operations forces
SOFPARS	special operations forces planning and rehearsal system
SOLE	special operations liaison element
SOTF	special operations task force
SOWT	special operations weather team
TIP	target intelligence package
TSOC	theater special operations command
U.S.	United States
USAJFKSWCS	United States Army John F. Kennedy Special Warfare Center and School
USASOC	United States Army Special Operations Command
USSOCOM	United States Special Operations Command

SECTION II – TERMS

Army special operations forces

Those Active and Reserve Component Army forces designated by the Secretary of Defense that are specifically organized, trained, and equipped to conduct and support special operations. Also called **ARSOF**. (FM 3-05)

joint force special operations component commander

The commander within a unified command, subordinate unified command, or joint task force responsible to the establishing commander for making recommendations on the proper employment of assigned, attached, and/or made available for tasking special operations forces and assets; planning and coordinating special operations; or accomplishing such operational missions as may be assigned. The joint force special operations component commander is given the authority necessary to accomplish missions and tasks assigned by the establishing commander. Also called **JFSOCC**. (JP 3-0)

joint special operations air component commander

The commander within a joint force special operations command responsible for planning and executing joint special operations air activities. Also called **JSOACC**. (JP 3-05)

joint special operations area

An area of land, sea, and airspace assigned by a joint force commander to the commander of a joint special operations force to conduct special operations activities. It may be limited in size to accommodate a discrete direct action mission or may be extensive enough to allow a continuing broad range of unconventional warfare operations. Also called **JSOA**. (JP 3-0)

joint special operations task force

A joint task force composed of special operations units from more than one Service, formed to carry out a specific special operation or prosecute special operations in support of a theater campaign or other operations. Also called **JSOTF**. (JP 3-05)

* **special air operation**

An air operation conducted in support of special operations and other clandestine, covert, and military information support activities.

special operations

Operations requiring unique modes of employment, tactical techniques, equipment and training often conducted in hostile, denied, or politically sensitive environments and characterized by one or more of the following: time sensitive, clandestine, low visibility, conducted with and/or through indigenous forces, requiring regional expertise, and/or a high degree of risk. Also called **SO**. (JP 3-05)

special operations forces

Those Active and Reserve Component forces of the Military Services designated by the Secretary of Defense and specifically organized, trained, and equipped to conduct and support special operations. Also called **SOF**. (JP 3-05)

special operations liaison element

A special operations liaison team provided by the joint force special operations component commander to the joint force air component commander (if designated), or appropriate Service component air command and control organization, to coordinate, deconflict, and integrate special operations air, surface, and subsurface operations with conventional air operations. Also called **SOLE**. (JP 3-05)

special operations task force

A temporary or semipermanent grouping of ARSOF units under one commander and formed to carry out a specific operation or a continuing mission. Also called **SOTF**. (FM 3-05)

This page intentionally left blank.

References

SOURCES USED
These are the sources quoted or paraphrased in this publication.

Army Forms
Department of the Army Forms are available on the Army Publishing Directorate web site (www.apd.army.mil).
DA Form 2028 (Recommended Changes to Publications and Blank Forms).

Army Publications
AR 95-1, *Flight Regulations*, 12 November 2008.
ATTP 4-10, *Operational Contract Support Tactics, Techniques, and Procedures*, 20 June 2011.
FM 3-04.126, *Attack Reconnaissance Helicopter Operations*, 16 February 2007.
FM 3-05, *Army Special Operations Forces*, 1 December 2010.
FM 3-05.30, *Psychological Operations*, 15 April 2005.
FM 3-05.40, *Civil Affairs Operations*, 29 September 2006.
FM 3-05.140, *Army Special Operations Forces Logistics*, 12 February 2009.
FM 3-05.160, *Army Special Operations Forces Communications System*, 15 October 2009.
FM 3-09.34, *Multi-Service Tactics, Techniques, and Procedures for Kill Box Employment*, 4 August 2009.
FM 4-0, *Sustainment*, 30 April 2009.
FM 5-0, *The Operations Process*, 26 March 2010.
TC 3-04.93, *Aeromedical Training for Flight Personnel*, 31 August 2009.

Department of Defense Publications
DOD Directive 4270.5, *Military Construction*, 12 February 2005.

Joint Publications
JP 1-02, *Department of Defense Dictionary of Military and Associated Terms*, 8 November 2010.
JP 3-0, *Joint Operations*, 11 August 2011.
JP 3-05, *Special Operations*, 18 April 2011.
JP 3-08, *Interorganizational Coordination During Joint Operations*, 24 June 2011.
JP 3-09.3, *Close Air Support*, 8 July 2009.
JP 3-35, *Deployment and Redeployment Operations*, 7 May 2007.
JP 4-0, *Joint Logistics*, 18 July 2008.
JP 5-0, *Joint Operation Planning*, 11 August 2011.

DOCUMENTS NEEDED
None.

READINGS RECOMMENDED
None.

This page intentionally left blank.

Index

A

aircraft capabilities matrix, B-29 through B-31
airdrop, 7-2, 7-6, A-19, A-28
airspace command and control, 2-5, 2-6
airspace control measure, 2-5, 2-6
airspace control order, 2-4, 2-5
airspace deconfliction, 2-5, A-11
assets and techniques, 6-2
attack helicopter, 1-5, 1-7, 4-8, 6-2, A-3, A-8, A-28, B-4, B-5, B-13
aviation life support system, 7-2, 7-3, 7-5, A-11, A-12

B

billeting, 7-2, 7-8, A-12

C

Civil Affairs, 1-3, A-25
close air support, 1-3, 1-5 through 1-7, 3-3, 4-9, 5-5, A-3, B-5, B-13
close combat attack, 1-3, 1-5 through 1-7, 3-3, 4-9, A-3, B-13
combat search and rescue, 1-6 1-8, 4-7, A-9, B-10, B-11, B-17, B-24
control authority, 2-5, 2-6, 5-3
counterintelligence, 1-8, 4-1, 4-4, 4-5, A-16, A-17
countermobility, 7-6
counterterrorism, 1-3, 1-4

D

defensive armed penetrator (DAP), 2-4, 7-6, A-8, B-10, B-12 through B-16, B-29, B-30
developed and undeveloped theaters, 7-4, 7-6
direct action, 1-3, 1-5, 2-8, 4-7, 4-9, 6-2, A-15, A-16, A-19, A-20, B-13

E

engineer support, 7-6

evasion and recovery, 1-6, 1-8, 2-8

F

field artillery, A-3
finance, 7-6, A-12
fire support, 2-4, 2-6, 2-8, 3-3, 6-1, 6-2, A-3, A-28, A-29, B-4, B-10, B-11
food service, 7-3, 7-8
force health protection, 7-3, 7-5
foreign internal defense (FID), 1-2, 1-3, 2-8, A-15, A-20, A-22
forward arming and refueling point, 2-6, 2-8, 7-2, 7-3, 7-6, A-2, A-17, A-26, B-24, B-29
forward-looking infrared, A-6, B-2, B-5, B-13, B-15, B-17, B-27, B-31
fratricide, 1-8, 2-4, 2-5, 2-8, 3-4, 3-6, 6-3, A-3, A-8, A-16, A-18
funding, 7-6

H

host nation, 1-2, 3-1, 3-3, 4-4, 7-8

I

information operations, 1-4

J

joint suppression of enemy air defenses, 6-1

L

landing zone, 1-2, 1-3, 3-4, 4-1, 4-3, 4-5 through 4-8, 7-6, A-2, A-3, A-7 through A-10, A-17, A-19, A-20, A-22, A-26, A-28, B-16
liaison officer, 1-6, 2-7, 2-8, 3-5, 3-6, 5-3
limitations, v, 2-7, 4-7, 7-3, B-5
logistics planning, 7-1, 7-2

M

maritime operations, 1-6
Military Information Support operations (MISO), 1-3, B-13, B-20

mission planning folder, 2-1, 4-1, A-1, A-15
MQ-1C (Grey Eagle), 4-9 through 4-11, 5-5, 5-6

N

night vision device, B-26, B-28, B-29
night vision goggle, 3-4, 4-7, A-11, B-3, B-6, B-14, B-16, B-21, B-28

O

operation order, 2-1, 5-1, A-1 through A-15
operational considerations, v, 3-4

P

principles of war, 2-2, 2-3

R

reconnaissance, 1-6, 1-7, 2-4, 3-1, 4-5, 4-9, 4-10, 6-2, 7-6, A-8, B-1, B-6
redeployment, 3-1 through 3-3
RQ-7B (Shadow), 4-9 through 4-11, 5-5, 5-7
RQ-11B (Raven) 4-9 through 4-11, 5-8
rules of engagement, 5-5, A-8, A-10, A-18, A-25, A-27

S

search and rescue, 2-4, 2-6, 5-3, 7-3, A-9, A-14, A-27
special reconnaissance, 1-3, 1-6, 1-7, 2-8, 4-7, 4-9, A-15, A-16, A-19, A-20
statement of requirement, 7-1, 7-3 through 7-5
suppression of enemy air defenses, 2-8, 6-1, A-3, A-10
survivability, 2-4, 2-7, 4-1, 4-3, 4-7, A-3, A-5, A-6, A-8, A-11, A-12, B-2, B-5, B-10, B-11, B-17 through B-19, B-25
sustainment, v, 1-1, 1-5, 2-6, 3-2, 3-3, 3-5, 4-8, 7-1 through 7-7, A-15, A-16, A-27

Index

Sustainment Brigade (Special Operations) (Airborne) (SB[SO][A]), 3-2, 3-5, 7-1, 7-2, 7-4 through 7-7

T

terrain, 1-6, 1-8, 2-1, 2-6, 3-3, 3-4, 4-1, 4-3, 4-6 through 4-8, 7-4, 7-6, A-1, A-10, A-16, A-18, A-19, B-1, B-2, B-5, B-8, B-15, B-16, B-20, B-25 through B-27

U

unmanned aircraft system, 4-5, 4-9, 4-10, 5-5, 5-6

W

weapons of mass destruction, 1-3

weather support, 3-4, 4-1, 4-5

FM 3-76 (FM 3-05.60)
28 October 2011

By Order of the Secretary of the Army:

RAYMOND T. ODIERNO
General, United States Army
Chief of Staff

Official:

JOYCE E. MORROW
Administrative Assistant to the
Secretary of the Army
1122302

DISTRIBUTION:

Active Army, Army National Guard, and U.S. Army Reserve: To be distributed in accordance with initial distribution number 114860, requirements for FM 3-76.

PIN: 102269-000

www.ingramcontent.com/pod-product-compliance
Lightning Source LLC
Chambersburg PA
CBHW050104230526
45470CB00004B/1676